Energy, Environment,
Populations, and Food

P9-CEO-526

Energy, Environment, Populations, and Food

Our Four Interdependent Crises

George L. Tuve

A Wiley-Interscience Publication

JOHN WILEY & SONS

New York / London / Sydney / Toronto

Library of Congress Cataloging in Publication Data

Tuve, George Lewis.
    Energy, environment, populations, and food.

    "A Wiley-Interscience publication."
    Includes bibliographical references and index.
    1. Energy policy—United States. 2. Environmental policy—United States. 3. Food supply. 4. Population. 5. Economic forecasting. I. Title.
HD9502.U52T89   301.31'0973   76-40351
ISBN 0-471-02091-5
ISBN 0-471-02090-7 pbk.

Printed in the United States of America

10 9 8 7 6 5 4 3 2 1

Preface

A succession of developments in the 1970s has called attention to several economic and technical influences that are producing decided changes in the American life-style. The most conspicuous of these have been the energy crisis and the growing demand that environmental pollution be brought under control. The Russian wheat deals point to some of the causes of the food price spiral and of inflation in general, and also remind us that only a small proportion of the human race is as well fed as we are. Waning United States aid to developing nations has been attacked and advocates of zero population growth emphasize that famines are imminent. High prices of several essential metals have been experienced, stimulating an expectation of wide future shortages of many industrial materials, and of the slowing of economic growth.

Each of these problems has been widely discussed in recent books and authoritative reports. Short summaries are repeated in magazines, newspapers, and broadcasts. Enthusiasm for one particular remedy or another tends to becloud the fact that these developments are all interrelated, and that a favorite solution for one may make the others worse. This book emphasizes their interdependence and the importance of a systems approach or holistic viewpoint.

The futurist tends to emphasize a present defect and to prescribe an ideal solution. Business and government idealize rapid economic growth. This author insists that unlimited growth and limited resources are incompatible, and this book explores the limitations imposed by capital supply and manpower when the four problems of energy, environment, populations, and food are jointly considered.

Emphasis here is on compatible objectives and the practical means for attaining them. Future planning, whether by the individual, business, or government, needs the firm guidance of past experience. This small book, while making no pretense of completeness, does summarize in detail and in quantitative terms the past experience, present status, and future prospects in each of the four interrelated fields indicated by its title. This examination is supported by an adequate collection of official statistics pertinent to each field (see Appendix).

Many specialized studies are quoted and referenced, the available data are carefully analyzed, but the story is kept simple and the future is viewed with confident optimism. Most references are to summarizing reports of official agencies or of major investigations. Conclusions from quantitative data are included in the text, but here there is room for opinion and some viewpoints have no doubt been slighted.

Most Americans will agree that such urgent improvements as environmental protection, the development of adequate energy resources, and better management of our great contributions to the world food supply should be planned and executed within our system of private enterprise. Some government regulation is obviously necessary, and some government participation in the research and planning phases is dictated by the national interests of conservation of resources, of defense, and of international cooperation in assisting the developing nations. The discussion and future projections herein are presented from the viewpoint of full confidence in our present system and its future evolution. Great personal opportunities are seen, and the challenges can be most smoothly met by full utilization of our existing systems of business and government.

Our complex and technical society has many other problems not discussed here. Successively they loom into importance and demand immediate attention from the public, from the business world, from Congress and the Administration. Remedies and adjustments are then initiated, the emergency subsides, and the public forgets about it. An oil embargo, a sharp increase in meat or sugar prices, a threatened famine, or a local war—these focus attention and emergency action. This is too often our method. But the problems persist because change is slow. Long-term solutions are called for and an educational approach is necessary. In this context the student, the planner, the executive, and the legislator all have need for facts and background. This book attempts to fill the need, presenting the material in quantitative terms, and arranged for easy reference. It deals with future opportunities and the limitations and risks, but predicts slower future growth.

As to the educational approach, most colleges and many high schools are rapidly adding to their coverage in these several fields of future planning. The individual courses differ greatly, but interdisciplinary approaches are increasing. The crises in energy supplies and environmental protection, the problems of overpopulation and of food supply, the great need for adequate diets and for economic assistance for the underprivileged, the excessive demands of unlimited growth, all have become imminent concerns in the preparation for a life career.

GEORGE L. TUVE

Cleveland, Ohio
July 1976

Contents

Tables

Page numbers in parentheses are text references.

Energy, Environment, Populations, and Food

INTRODUCTION

Four great challenges of the final quarter of the twentieth century were hardly anticipated. Today's crises in the fields of energy supplies, environmental control, population crowding, and shortages of food and materials are closely interrelated. It is gradually being realized that they demand a holistic or systems approach.

The enthusiastic mass production that has long continued in several parallel segments of the U.S. economy, notably the energy industries and agriculture, was stimulated by the broad expanse of this nation, its resources and its markets. There has also been a century-long inventive expansion of transportation and communication and a startling growth in population.

Views of earth from space emphasized the fact that the planet and its bounties are by no means infinite. Crowding, pollution, hunger, and shortages have demonstrated that there are limits to growth. While these have been local ills for centuries they have always been temporary or curable. Now, especially since 1950, there have been many well-supported warnings that such ills are reaching crisis proportions and will not disappear.

The curving graphs of exponential growth of populations, production, and resource consumption have a way of reaching upward toward infinity even when the annual rate of growth seems small. This is well known, but the fact that these three kinds of growth comprise an interrelated system has been largely overlooked. Now we see that a 5% yearly increase in the use of energy not only points to fuel depletion, but is also a main cause for the growth of pollution. Growth in production of the automobile, the personal energy machine, generates a simultaneous growth in the rest of the economy, but the entire process depends on increases in population and food supply. Petroleum is our best fuel, but its accompanying natural gas is an invaluable raw material for making fertilizer. For mass production on the farm, more and more petroleum fuel is being used per calorie of food produced.

Increasing populations of living things are at the root of the crises in energy, environment, food, and many basic materials. In much of the underdeveloped world the 3% yearly increase in people to be fed creates an almost hopeless prospect. Even the improved food crops and methods can never keep up with an unrestricted expansion of populations of humans, animals, pets, rodents, birds, and insects. The means for population control are available, and humans must be educated to use them.

Affluence in the developed countries adds to food shortages. In poorer countries food is mostly cereals and fish. But a pound of beef requires 8 pounds or more of cereals as cattle feed, adds a waste problem, and uses more land, plus power and refrigeration. Poultry farms and cattle feedlots consume great quantities of fish and oil that might well serve as human food.

The 1970s can yet become the period during which the four interacting crises are finally recognized as a single broad challenge and placed in the center of the political arena. Here is a great opportunity for effective improvement and cure. The disbelief, inaction, and argument, must be dispelled. It is based largely on ignorance of facts and convictions of futility. This calls for education, information, and communication. Disagreements like those between the president and the Congress can and must be resolved.

Farseeing leaders in government, in the UN, the OECD, and in education and business are urging intensive research programs and wide systems approaches. Universities and teachers find a quick response from the younger generation, and an eagerness for information and participation. This book is intended as a small contribution to this pressing opportunity. Its first objective is to provide a compact text and a source of quick reference on past experience, present status, and future proposals in the four interdependent areas. Authentic summaries of facts to the present and outlines of best thinking on future plans are needed as a basis for action by leaders and support by followers.

After the examination of our resources and accomplishments to date and of proposals for our future development, the last four chapters in this book suggest several practical limitations to the carrying out of our future plans. Our role in world development is first examined with reference to foreign aid to underdeveloped countries. Throughout the book, the focus is principally on the U.S. and the next 25 years, but it is emphasized that wise planning requires the world's richest nation to divert substantial resources to the development of have-not economies, where the pressures of population growth and food shortage are already ominous.

The American ideal of maximum economic growth will be very difficult to abandon, but Chapters VI and VII contrast a necessary *slow future growth* with the ups and downs of the past. Too many ambitious single-purpose plans are made with little regard to the interdependent effects. Each field actually competes with all the others for capital, manpower, and political approval. Can we become self-sufficient in energy, fully protect our environment, help feed the world, and assist in economic developments, all at the same time, and yet slow our growth and retreat to a smaller and fairer share of the world's raw materials?

Adjusting to these demands will be a real struggle, and it promises some changes in the U.S. life-style. The final chapter herewith offers several predictions about this adjustment—controversial perhaps, but largely optimistic.

Chapter One

Welcome to the Energy Crisis

A. OUR FIRST ACUTE EXPERIENCE

When the Arabs shut off U.S. oil imports in 1973 and then quadrupled the prices, they did us a real favor. Everyone was affected by the embargo and still remembers it. Suddenly the prospect of shortages of oil and natural gas changed from a problem for the experts (who had been warning us of it for years) to a reality for every person who drives a car or lives where it is cold in winter. The energy crisis was an acute experience, and the fact that it affected everyone led us to do some much needed thinking. Was this a short-time emergency or might it fore-shadow other difficulties that could result from our ideal of continuous growth? We soon learned that, although the United States contains little more than one-twentieth of the world's population, we use one-third of the world's annual fuel supply. A similar ratio applies to our consumption of many other critical materials, including several essential metals. It is hardly strange that we have been criticized for this.

Our amazing economic growth since World War II has required a tremendous flow of raw materials, but it has also greatly increased the pollution of our natural environment. High rates of energy use are clearly a major cause of pollution, much of it concentrated wherever fuel is burned, and large investments will be necessary for much improvement in this respect.

Only after our own energy crisis in 1973–1974 did we begin to realize that the crisis had become worldwide and was much more acute in countries that have little or no fuel sources within their own borders. After all, the United States is one of the world's leading oil producers, while Japan, for example, has no domestic oil. Therefore her entire economy faced a shutdown. India, with three times the U.S. population, produces no oil either. As in many other underdeveloped countries, India's population is growing so fast that the food supply cannot keep up. They are already short of food, and their supply must be increased by one-third every decade. The oil embargo paralyzed much of their food transport and also threatened their modest industries. Hunger and starvation soon appeared in many such countries, and they turned to the United States and Canada for help. Thus the oil crisis affected the farmer here at home directly, because our modern agriculture is highly energy-intensive, but also indirectly because of a large foreign demand for our grain. Soon this contributed to higher U.S. food prices.

Today, thanks largely to the energy crisis, every American is aware of the attention being paid in the public press to all four of the subjects named in the title of this book: energy, environment, populations, and food. These topics also appear in weekly news magazines and television documentaries. Serious students now realize that difficulties of near-crisis proportion face our country and its citizens in each of these fields. It is often overlooked, however, that the four crises are highly interdependent.

The experience of the energy crisis has given the U.S. citizen a new perspective that would otherwise have taken years to acquire. A new set of objectives is arising to replace the post-World War I aims built on the ideal of unlimited growth. The phenomena of growth provide a basic and reasonable connection between shortages of energy and of many basic materials, pollution and ecology problems, high food prices, crowded city traffic, congested recreation areas, and many other disturbing features of the 1970s. It now becomes evident that not enough attention has been paid to ecology enthusiasts, energy planners, foreign-aid committees, and planned parenthood and zero population growth (ZPG) organizations. Each group has made great strides toward improvement, but we could do so much better if the basic conflict were always recognized, namely, boundless growth versus limited resources and room. The capacities of spaceship earth are not unlimited. Shortages are appearing, and wastes pollute the environment. Our use of the earth's available resources is not highly efficient and effective either, for reasons that are both human and economic. We must understand the basic limitations and gradually change our life-style accordingly. The challenges demanding this adjustment are greater for Americans than for anyone else, because both our resources and our use of them have been the largest.

The United States has some of the best farmland in the world and the most highly developed agricultural technology for quantity production. Only 5% of our workers are needed on farms, yet we can raise crops vastly in excess of what we need for ourselves. We have large forests, a great mining industry, massive amounts of coal, petroleum, and natural gas, and ample transportation. The prerecession U.S. gross national product (GNP) was about $5000 per capita, while in the poorer countries it was closer to $200. As we look forward to scarcities and slower growth, the accompanying changes will be great, but the greatest future shock will be to our philosophy and our expectations.

America is now challenged to abandon *quantity* as a foremost goal. The worldwide energy crisis caused little more than inconvenience in the United States, but its effects in many poorer countries have been drastic. Their daily food supply is not much more than one-half of ours, even though most of their workers are employed in agriculture. Contrary to hopes for a "green revolution" there was no increase in food production per capita in Africa or East Asia in the 12 years 1961–1972, because their populations grew so fast. When costs suddenly doubled for transportation, fertilizer, lighting, and power, it almost threatened the existence of such nations. They have no cushion of individual resources to meet these demands. This is not easy for us to realize.

The oil embargo was a great educational force. Vital knowledge was spread from a few specialists to almost everyone. Very gradually the public is becoming aware that energy shortages, pollution, and world shortages of food and materials are all interrelated and that remedial action in one of these fields must take into account the secondary effects in each of the others.

A great many people are not yet convinced, however. The greatest hope for eventual effective action lies in systematic educational presentation of the facts, underlying causes, and future prospects. Fortunately, colleges and schools have undertaken this in a big way. Hundreds of colleges and thousands of teachers are treating these problems, from junior high schools to postgraduate courses at universities. The main focus of attention from course to course depends largely on the teacher. It may be the environment and its control, futuristics and future planning, energy engineering, or demography and population growth. The economic and political science aspects of the management of resources are often emphasized. Whatever the emphasis, the interdependencies are important.

The first objective of this book is to describe briefly our past experience in the four fields mentioned in its title, documenting that outline by pertinent statistics (see Appendix). Prospects for the future in each field are examined in the light of where we stand today, and within the limitations dictated by past experience. In these discussions heavy reliance is placed on the conclusions drawn by the many specialists who have studied each field. Reports to the president and to Congress, studies commissioned by foundations, and proposals by research centers and university experts are freely quoted and compared. Special attention is given to such interdependence, for example, as the probable effects on the environment of the various programs to increase the energy supply.

The worldwide aspects of the critical problems in the four fields become a part of the discussion, with particular reference to proposals for action in the near future. Some of the greatest difficulties are concerned with the less developed countries, and a full chapter is devoted to U.S. experience in foreign aid and the bilateral and international activities related thereto.

Perhaps the most significant lesson learned from the energy shortage has been that *growth* is a basic causative factor—population growth and economic growth in particular. Throughout this book, rates of growth are computed and tabulated. The "limits to growth" pointed out by the Club of Rome and others, are emphasized, and the general conclusion is that the future dictates slower rates of growth. In this context, probable future developments in the United States are explored, with due regard to where we are now, that is, with due consideration of the recession and the inflation of the mid-1970s.

Finally, the last three chapters explore the contrasts between the past and the probable future of the U.S. economy and the accompanying changes in our style of life. What are the pressures for change, the risks, and the future opportunities? What is the proper role of government as compared with that of private enterprise? The "art of the possible" must obviously control future efforts to emerge from crisis situations. Political obstacles are awesome, but plans are also limited by available manpower and by the capital required for their accomplishment. Nevertheless, full confidence is expressed in the ability of our economic and political systems to adjust to the new demands. I insist that in spite of expected changes in life-style and viewpoint, the future opportunities, both national and individual, are bright indeed.

B.   A NEW AMERICAN LIFE-STYLE?

It is not wholly accurate to describe present and prospective conditions in the United States as "new." Immediately after World War I Americans were warned about limited resources of petroleum and natural gas.

Geologists, engineers in the energy industries, and government specialists such as the Geological Survey and the Bureau of Mines presented figures and discussed the dangerous prospects of limited U.S. supplies. Pollution warnings were common as early as the turn of the century. Chicago, Pittsburgh, and other cities commissioned large volumes of reports on their smoke and air pollution problems. Water purity, too, has long been a subject of whole shelves of government reports. Theodore Roosevelt campaigned for conservation. Since Hiroshima the hazards of nuclear radiation have become a major concern. In fact, the government was worried about it from the start of the Manhattan Project. Warnings about food supplies and the necessity for birth control have been in evidence ever since Malthus (1798). Depression has often followed runaway spending and debt. Why so little action in response to these warnings?

Actually, the specific improvements have been great. The overall fuel efficiency of steam-electric plants was doubled. The air was cleaned up in Chicago and Pittsburgh. Every U.S. state, town, and hamlet set up pure water standards. Radiation hazards are now subject to stringent national and international limitation. We have been sharing our food crops with other nations. In the fields of U.S. production of energy and food there has been great progress. New discoveries, improvements in technology and quantity production methods, plus an amazing growth on the part of suppliers, has kept up with increases in demand, until recently. Better recovery from the oil well, improved refining, vast new oil and gas fields, and excellent distribution made energy optimists of all but the very few who were aware of the narrow margins. In agriculture new implements, the combine, chemical fertilizers, irrigation, new seeds and hybrids, crop drying and improved storage, and great ranches, more than met the nation's demands. Long-range progress has been phenomenal in spite of a few serious setbacks. But now several things have changed, and there are famine and malnutrition in many lands and a world crisis in energy supplies plus many threats of economic disorder.

What are the new demands? What must we do differently? The answer is that everyone must change: the producer, the distributor, the government, the consumer, the attitude of the public. Local and partial remedies and furious production to meet increased demands and a growth philosophy of "the sky is the limit" must give way to integrated plans and full public information and discussion. Business and technology must turn in new directions, away from the ideal of quantity growth and toward an emphasis on quality, toward service, long-lived products, conservation of materials, and economy and efficiency in the use of energy. Thoughtless waste of materials and energy is no longer tolerable. Conservation must become a guiding objective. Emphasis will be placed on benefits for customers and employees, and sensible adjustments must be made to long-time expectations of slow overall growth even after full recovery from the recession. These are major demands, but they are stimulating challenges. They will not be readily accepted. The changes will be slow, except for emergencies, and they will be resisted by the marketplace, by organized groups such as unions, by government officeholders, and by office seekers. Such complete changes in objectives and philosophy are not easily made, and there will be many holdouts. Here again we are being taught by the energy crisis. The two grave questions are: Can the new requirements be met without disruption and violence? What can be accomplished *soon*, in the next 10 or 25 years? Our democratic processes are slow, and international cooperation will be needed too.

Underlying much of the difficulty is the threatening fact of vast population growth. World population has more than doubled in a lifetime. Present growth rates cannot be tolerated in the future. The postwar "baby bulge" added greatly to the energy crisis. The awesome "Limits to Growth" as projected by the Club of Rome focused world attention on the importance of population control. A startling problem is posed by the fact that the underdeveloped countries are growing more than twice as fast as we are. The most intensive programs of conservation, pollution control, and protection of living standards will fail unless population growth can be restrained. Will it be possible to carry out a workable and coordinated program to meet these challenges? If this is not accomplished soon, violence and disorganization may result. We do not accept doomsday prophesies, however. Humans are adaptable, and many options are open.

The shock treatment administered by the 1973 oil embargo has been successful. Now we realize that with "Mankind at the Turning Point" (see footnote 3), and the interdependent problems clearly defined in terms of four crises, energy, environment, populations, and food, the next step is to obtain the information and make future plans accordingly. With the facts of previous experience as a guide, every citizen can evaluate the alternatives and has the opportunity to contribute.

The twentieth century has been an era of rewarding economic change, but at no time have the rewards been more promising than in the late 1970s. Moreover, the situation is unique in that no individual is denied a part in the action—unless he or she is quite incapable of change.

Chapter Two

The Capacity
Limits of
Spaceship Earth

A. WORLD RESOURCES VERSUS POPULATION

The Population Explosion is a fairly recent phenomenon. Its threats of shortages and famine are not yet fully appreciated by most people in the Western countries for one very good reason, namely, the idea has been accepted that an "economy of abundance" has been reached in the twentieth century. History as studied in our schools describes a continual struggle of earlier peoples to subsist on the resources available to them and tells of the endless wars that have been fought to gain the assets of new lands. It is not surprising that, after Western countries developed the new marvels of production and distribution, everyone in those countries was willing to accept the idea that the long struggles against scarcity and want had been won. Of course, the spread of this conviction of plenty was disturbed somewhat by the great depression of the 1930s and the shortages during the war that soon followed. But a quarter-century of uninterrupted good times for the industrialized nations again strengthened the convictions of abundance. The older members of the population could hardly believe it. They knew too much about depression hardship and about the struggles of their parents and grandparents. But as generations matured who had known nothing but prosperity, the acceptance of a philosophy of abundance became almost complete.

Such satisfying convictions change very slowly, even though an occasional recession does modify them. U.S. shortages and high prices today, as those of motor fuel and many foods, are very likely to be seen as conspiracies of profiteers. There has been just enough truth in stories of excessive profits to counteract any great interest by the general public in the new pronouncements about dwindling resources.

In the 1950s several books appeared which warned of the urgent need to control the growth of population and called attention to the deplorable conditions of overpopulation and impending food shortage in many of the less developed countries. In the 1960s the environmental movement became important, with its sharp emphasis on reducing the pollution of air, water, and land in highly developed and populated areas. In the 1970s came the energy crisis. The whole picture is finally being put together, with a realization that mankind does *not* have unlimited resources on the earth. A new perspective was furnished by the space program. Detached observers advanced the startling idea that the earth, a finite ship in a great universe, has decidedly fixed and limited resources. What, then, can come from a seemingly unlimited expansion of the earth's population? As a result of these developments, many studies began to be made, matching available resources against existing and expanding populations.

In several of the nonindustrialized countries, the present annual rate of population growth is about 3.5% per year; and for the entire world the population growth is almost 2% per year. The arithmetical parallel between population growth rates and compound interest is quickly realized by our business civilization; and from the interest tables it is seen that money, or population, if it has a compound growth rate of 3.5% per year, increases about thirty-two times in 100 years. If the rate is 2%, the corresponding increase is about $7\frac{1}{4}$ times. Using this latter multiplier we have a prospect of a world population of almost 30 billion before the year 2080. This is ridiculous, since the earth's resources could not support that many by methods now known; but this simple projection certainly indicates that population control is an urgent problem. In a country growing at a 3.5% rate, the population doubles in 20 years, which means that *every* generation must at least double the food supply. In a country still using primitive methods in agriculture, and poorly fed even now, an assignment like this is quite impossible to meet. Such a rapid and continuing growth in food supply could not even be reached in a country that has an advanced agricultural economy.

Worldwide data on population and resources are presented in condensed form, country by country, in Table 1.[1] This table shows the rate of

[1] Table 1.   National Populations and Their Living Standards, Year 1970: Countries of Over One Million Population.

population growth in each country and the extent of affluence or poverty; and it gives several indexes of living standards. Careful review of these statistics shows that the plight of the underdeveloped countries contrasts pitifully with the style of life in the industrialized world built on abundance. The UN, after a study of extensive data for the year 1970, estimated that the average "gross domestic product for the developed market economies" was $2960 per capita; for the developing market economies, it was only $220; and for east and southeast Asia (excluding Japan), the gross domestic product averaged $120 per capita in 1969.

Most products mined from the earth to supply the world's increasing populations are nonrenewable. Prospects of adequate reclaiming of used materials are dim. The special case of fossil fuels and energy sources is examined in the next chapter, and the rates of use and apparent reserves of many useful metals are presented in Table 2.[2] It is very evident that within the lifetimes of the present U.S. population some serious adjustments must be faced. For example, much of our wide use of iron and steel depends either on the protection of this base metal from corrosion by means of coatings, or on the improvement of its properties by alloying. Supplies of the two most common coating metals, tin and zinc, are threatened; and the same is true for three important alloying metals: manganese, chromium, and cobalt. Many of the coatings that are nonmetallic are based on petroleum, another limited resource.

The United States has practically no native ores for the production of aluminum, chromium, cobalt, manganese, nickel, platinum, and tin. We have less than 6% of the world's reserves of gold, mercury, titanium, and tungsten (Table 2). Considering that the U.S. population is about 5% of the world's total, we consume over four times our proportionate share of sixteen of the nineteen metals listed in Table 2; and for at least twelve important metals, our imports exceed domestic production. Yet our total consumption of nonferrous metals more than doubled from 1960 to 1973, while population increased only 17%.

What can be expected as to the continuing and increasing supplies from natural resources for the fast-growing human population? It has become a habit to expect that discovery and production of these supplies can and will expand as needed. However, some critical resources are now seen to be strictly limited. Among these are all widely used forms of energy, several strategic metals and minerals, and certain waste reservoirs, including even the ocean and the atmosphere. However, many resources, both material and environmental, can be renewed by nature's processes, by substitution, or by inventive reclaiming and recycling. Much more attention should be paid to these possibilities.

[2] Table 2. U.S. Position in the World Metal Market: Production, Consumption and Reserves.

Several computer-based projections of the long-range effects of population growth have been published, and these should serve as an alert; but many people tend to discard these warnings with the assertion that the human race has always been adaptable and that such extreme pictures are not worthy of serious attention because the problems will be solved long before intolerable conditions can develop.[3] Actually, the expected adaptations, in the form of increasing attention to the possible effects of unlimited growth, already have a good start. Even so, several crises have developed, and the damage could well outrun the corrective efforts. Only with widespread understanding of root causes and applicable remedies can long-term corrective programs emerge. To develop this understanding we must first heed the facts about population growth and about food sources and the limits of food production. Then, since our modern life is so dependent on energy, we must explore energy sources, present and prospective, and the costs of developing substitutes for the very convenient present fuels, oil and natural gas. Are we actually nearing the capacity limits of the environment with respect to the ability of air, land, and oceans to absorb waste? If this is true, even in local situations, it imposes yet another limitation on population growth; and every proposal for increasing supplies of food, energy, and other human necessities must be examined for its environmental impact. America, as the world's largest producer of food for export and largest importer of oil and mineral wealth, has a special responsibility with regard to population growth.

Many people are not convinced that the world's capacity to sustain human and animal populations is so limited. They point to the vast open spaces that could be occupied. This possibility of spreading out must be conceded, although it is in direct conflict with the gregarious tendencies of the race. There have been many experiments in the forming of new communities remote or apart from existing occupied areas, but the process is slow and often unsuccessful. In the poorer countries especially, spreading the population increases the difficulties of distributing food, materials, and energy where transportation is already inadequate.

The development of natural resources in any country depends on the people available to do the job. In many countries with low living standards the capital required is almost unobtainable. In the United States, where natural resources have been overabundant and human resources have attained high development, material standards have

---

[3] Among the most impressive computer studies have been two reports to the Club of Rome: first, *The Limits to Growth*, by Dennis Meadows and associates, published in 1972, Potomac Associates, New York, and, second its sequel, *Mankind at the Turning Point*, by M. Mesarovic and E. Pestel, published in 1974, E. P. Dutton & Co., New York. See further discussion of these reports in Chapter 8, Section C.

reached such a high plateau that it is most unlikely that the developing nations can approach them within the foreseeable future. The poorer nations greatly resent this attitude, and they also resent some of our tactics to gain even further improvement for ourselves, particularly when they are furnishing much of the raw material essential to our industrial progress.

While most Western nations have come to the conclusion that world rates of population growth must be reduced because resources are limited, many of the developing nations insist that the affluent countries are selfishly advocating limits to population growth in order to protect their affluence. Only about one-fourth of the poorer nations have aggressive population control policies, and there is now a wave of hostility directed toward population control based on the accusation that the rich nations do not want the others to develop and become industrial competitors. Hostility was much in evidence at the UN Population Conference in 1974. If and when the great underprivileged majority of the world's people do become more affluent, it will doubtless put a strain on the world's resources; and the industrialized countries will no longer enjoy the lion's share.

B. THE GROWTH OF POPULATIONS

Few people realize how startling are the statistics of population growth. It took the human race thousands of years, up to about 1825, to reach a total living population of 1000 million, or 1 billion, people living on the earth. Only 100 years later, in the late 1920s, the total was two billion. Now, in another 50 years, a four billion total has been reached. All trends point to a world population of more than six billion by the year 2000.

A yearly world increase equal to all the people living in our five most populous states—California, New York, Texas, Pennsylvania, and Illinois—certainly calls for tremendous increases in world production of food, materials, and energy just to keep even, without any improvements anywhere. Almost two-thirds of the world's population is undernourished even now. Moreover, in many countries there just is not much room left; and population growth adds to malnutrition and misery. In India, for example, the population density is eight times what it is in the United States, and the area one-third as large; and they have a yearly population increase of almost 15 million, like adding to our eastern states each year a population equal to that of New England. What are the prospects of providing for this added population in India next year, and at the same time increasing the very low living standards of the 600 million already there? After all, their primitive farms average much less than 10 acres each, and their total national income per capita is about $100 per *year*. Several countries in Asia, and many in Africa, are no better off than India; and their populations are increasing even faster.

If present rates of world population growth are assumed to continue, it will be found by extrapolation that in another 600 years people would have 1 square yard apiece. While this is absurd, spaceship earth *does* have a limited capacity.

A whole library of books is available on demography and the control of growth of populations. The subject is covered here only to the extent of indicating the interrelation of population growth and the supplies of materials, food, energy, and living space available. It is the recent upward trend of population curves that is at the root of scarcities, malnutrition, pollution, and demands for more moderate life-styles, which are among the subjects covered in the following chapters.

It is an observed fact that as a nation becomes more prosperous and highly developed the birthrate goes down. In a few countries the annual population growth during the UN statistical period actually averaged very near zero, that is, ZPG had been attained. In a few other countries population growth is still needed, and there is no enthusiasm for a campaign to lower the birthrate. In view of the finite limits of world resources, however, it has become popular to clamor for ZPG; and the many ZPG organizations are calling for its early accomplishment. But there is a built-in potential for population increase in most countries (see Section G). In much of the world there was a "baby bulge" in population after World War II. Now, a generation later, this large younger group is in the midst of the reproductive period. Fair predictions of population growth can be calculated from existing census statistics, and the UN and national census bureaus have made extensive estimates.

Most countries now conduct careful census counts; but the dates of these enumerations often do not match, so the counts of total world or regional populations at a given time are less accurate. Some of the smaller nations with scattered populations have been able to make only approximate counts, and several countries do not publish their census data. The UN has made increasing efforts to collect and publish the best available population figures for all countries and to combine and to analyze the data by regions or groupings. It has also made projections of future population growth based on the existing age structure of the population in each country. Some of the UN data based on 1960 and 1970 census counts are given in Table 3,[4] and included are the corre-

---

[4] Table 3.   World Population and Its Growth, 1900–1985: UN Data and Projections, World Regions.

sponding projections for 1980 and 1985. The average cumulative yearly rates of growth are shown, as computed from tabulated population data. Similar but more detailed figures for the U.S. population are given in Table 4,[5] starting with 1800 and listing the minimum and maximum Census Bureau projections to the year 2000.

It will be noted that world population is expected to be almost five billion by 1985, having increased two billion in 25 years, and with no expectation that the rate of growth will subside in the coming 10 years. The growth rate in Europe, about three-fourths of 1%, is about the same as the 1972–1975 rate in the United States; but there is some question whether this low rate will continue in our country. Latin America and Africa have fast-growing populations; and during the 1970s they are expected to add 230 million people, as compared with about 56 million added in the United States and Europe combined. In the meantime Asia will gain ten times as many people as are added in the United States and Europe together. The almost three billion people in the developing countries, who cannot even now support themselves adequately, are expected to increase by 700 million in the 1970s; while the United States, Europe, and the USSR together add about 85 million to their 1970 population of less than a billion (see Chapter 5). ZPG is not likely to be attained in any of the areas or groups in this century.

While it is easy to cite population growth data that provide good scare headlines, it is much more important to understand the whole picture. The data indicate a present world population growth of close to 2% per year and several countries growing at the rate of 3% or more. Actually, the countries with a 3% growth rate constitute less than 10% of the world's population. At the other extreme, the population growth rate is less than 1% per year in countries with 15% of the world's population. Thus the world problem of population control is by no means hopeless, as it is mainly concerned with a small reduction among the 75% of the human race where population growth at present is only moderately excessive. A birth control program that would result in an average number of children per family that is only one less than the average now prevailing would go far toward solving the problem.

As to U.S. population growth during the balance of the century the best estimates are that the 1972–1975 rates of growth of about three-fourths of 1% are a minimum, and that growth will again increase and hover around 1% per year. Even this rate is much lower than that during any earlier period in our history except the depression years of the 1930s.

[5] Table 4. U.S. Population and Its Growth, 1800–2000: Total Growth and Average Yearly Growth.

While the estimates of future U.S. population growth are based primarily on the age components of the current female population, and on constant immigration, they could be upset by unexpected changes in economic or health conditions, by trends in family planning practices, or by variations in net immigration such as large admissions of refugees.

C. THE AMERICAN GROWTH PHILOSOPHY

The accepted goal for the enthusiastic advocates of population control is ZPG. They even talk about zero economic growth (ZEG). Opinion is widespread, however, that there is good reason to question *any* proposal for zero growth. Any no-growth concept seems foreign to all normal habits and thinking. Growth is the basis of life. It is safe to say that growth has always been a human ideal. Proposals for zero growth seem to contradict all past experience, even if they are limited to the fields of population and economic activity.

America has always expected continuous growth, ever since the first settlers. We aspire to personal growth and to growth of our family, assets, towns, industries, and country. We praise and marvel at nature's examples of growth—of crops, trees, flowers, and animals. We like to see growth in our business ventures, our schools and colleges, our clubs and professional societies, and our churches and charities.

Today's generations in particular have experienced growth beyond the wildest dreams of most of the world, past and present—growth in personal opportunities, personal possessions, and the good things of life. Charts of growth fill magazines and textbooks. Business news always features growth data, growth of sales, growth of assets, growth of profits, and growth of the GNP. Someone has called economists the "gurus of growth." (For a no-growth society, all economists must be retrained.) America not only lives by a philosophy of growth, she thinks her people are *entitled* to the amazing rates of growth seen in our lifetimes, in spite of any shortages in supply.

A hundred years ago, as the country grew, we needed more people to fill the land, to develop industries and, more recently, to provide the many services that go with a high standard of living. The problem is no longer that of filling forty-eight or fifty states. Now fewer and fewer farmers are needed, and people want the advantages of urban living, plus the privilege of going anywhere anytime. We want to be comfortable at all times, wherever we are; and we want to be independent, each choosing his or her own activities, and with more and more leisure time to enjoy them.

Things are different in the developing nations, which constitute two-thirds of the world's population. In most countries in which populations are growing faster than ours, the great percentage of the employed, except those in government, are engaged in agriculture. The United States has made the "demographic transition," so that few workers are needed for providing food; and most of the working population are engaged in industry, education, the service professions, and cultural development. How did the industrial nations make this transition from the agricultural pattern with its high birthrate, high death rate, and *low* standard of living, to the urbanized industrial pattern with its low birthrate, low death rate, and high standard of living? The development in each industrial country has been different, but the United States has been just about the most favored of all.

Economic growth statistics have long been preferred reading in America. Indicators of business growth are broken down into a dozen categories, and the final overall indicator is the GNP. The data in Table 5[6] give some indication of the results of the American growth philosophy. While the U.S. population doubled between 1920 and 1973, the current dollar value of the GNP increased fourteen times. College enrollments increased in a like ratio. In constant dollars, the personal income per capita more than quadrupled. This was the high-energy era, with electric power consumption increasing thirty times and automobiles almost fourteen times. The record, however, was in expenditures by governmental units.

The phenomenal U.S. growth since World War II deserves special attention. During the period from 1950 to 1973 the GNP per capita (corrected for inflation) doubled (Table 32). This was largely attained by the increase in productivity per manhour. Government expenditures increased much more rapidly, and the public debt per capita tripled (constant dollars). Most of the increase in public spending per capita was centered in the fields of benefits and services to individuals, that is, personal security, health, and education (Table 30). The inability of government tax income to keep up with these expenditures for personal welfare became even more serious after 1973.

In several fields the 1950–1973 U.S. growth was exponential, more rapid than a straight-line accumulation. The lines of increase were concave upward, that is the *rate* of growth kept increasing. Such growth cannot continue for long without radical adjustments, or even disaster. This is well illustrated by the present difficulties in energy supply and

[6] Table 5. U.S. Economy by Decades, 1900–1973: Personal Income, Energy, Cars, Education.

environmental pollution. This same limitation is basic to the world food problem.

By the late 1960s the yearly growth rates in the energy field, for example, were in the 4 to 8% range (Table 10). Such rates call for a doubling of the supply in 10 to 20 years, and the limits to such a process are evident. Neither resources, capital, nor manpower can sustain such rates of growth for a long period, and crisis conditions were to be expected. We have selected a yearly growth rate of about 2.5% as much more likely to be sustainable in the future (Table 19).

Readjustment experiences such as depression and unemployment in 1974–1976 were a shock to the American growth philosophy. In the struggle for resumption of "prosperity" the dangers of overcorrection are great. In certain areas high growth rates may be necessary to correct the deficiencies accumulated during the depression. But the ideal of maximum growth is still firmly implanted, and the great question is whether the public will accept growth rates that have tapered down to the 2 to 3% range, say, after 1980.

As soon as economic growth rate falters, there is much talk of a "recession"; and when some indexes level off to zero growth, a "depression" is declared. The business system does not manage well during a depression, nor does the government, and this causes panic. Overall, the marketplace resists any idea that zero growth, or even slow growth, can possibly or ever be a desirable condition. Thus the zero growth advocates are opposed at every turn, if not by reason, then by ridicule. Nevertheless, our troubles will soon make slow economic growth a permanent reality; and it is time to realize that effectiveness and efficiency are much more important than increase in size. In any business, once the critical or effective size has been attained, additional growth can even lower the internal efficiency.

In almost any joint human activity or organization, there is a critical size below which the operation is at a disadvantage. In a club, church, town, school, or local business, the activities and the facilities both remain inadequate until a certain size is attained, that is, until enough trained people are involved to carry out the objectives satisfactorily. For a business, this critical size depends largely on the size of the potential market. A great number of U.S. corporations are so large and so diversified that further growth is not necessary in order to increase their effectiveness. While their new lines are growing, some of the older ones approach obsolescence or unprofitability and should be retired. (Acquisitions and mergers for financial or tax manipulation are a different story, but they do not necessarily result in overall growth, nor in more effective operation.)

However, it is certain that the private enterprise system could not survive without lusty growth of individual units. Any small business, or specialist department in a large business, must struggle to grow rapidly until its critical size for maximum effectiveness is attained. It is not strange that the business community is slow to embrace any conviction that slower overall growth is either inevitable or desirable. Yet any businessperson could name many an example of a business that is highly progressive, successful, and profitable, but which has not increased in size for many years.

Unlimited growth is now the wrong objective. We are already aware of many organizations that have been so dedicated to continuous growth that they have become too cumbersome to accomplish their main objectives. This has even happened to whole industries.

It soon becomes evident when observing the effects of economic growth in the United States that these effects are intimately linked to conditions in other parts of the world. Not only the prices and supplies of gasoline and oil, but those of our food and raw materials, many of them imported, are related to the ills that follow unrestricted growth. In many cases the developing countries are involved. These countries comprise the majority of the human race, but they are not yet industrialized. Trade with them is intimately linked with foreign aid to the same countries, so the facts about American and worldwide aid programs must be explored (Chapter 5).

Population growth and industrial expansion have resulted in yet another related problem, pollution of the environment. Again, this is international, especially with respect to the atmosphere and the oceans. In our country the favorite method of dealing with these closely related effects of growth has been to treat each as a separate crisis and to apply immediate but separate remedies. It is time to modify the American growth philosophy and to apply an integrated approach, even an international attack. This is an immediate opportunity for long-term planning that will yield great benefits.

Our nation is a living organism, and it should be no surprise that exponential growth finally becomes cancerous, nor that the several phases of its growth are interdependent. One example of interdependence is the increased productivity in all industries that has been made possible by wide applications of energy. The use of energy has multiplied the powers of the individual. But sheer size and quantity are beginning to strain our resources and capabilities, decreasing our lebensraum and threatening our physical and social environment. The ideals of quality and service are being submerged in the race for quantity. While each symptom indicates the need for a special treatment, every such effort should become a part of an overall plan. A *systems approach* is essential, and the ideal of exponential growth must be questioned.

Recently there has been evidence of wider recognition that the goal of unlimited economic growth must be superceded by other objectives if the world's resources and its natural environment are to be adequate for all humankind, and for future generations. It is encouraging that late in 1974 the Organization for Economic Cooperation and Development (OECD), which was organized early in the 1960s with a purpose of working toward maximum economic growth, made a new declaration. This organization, of which practically all non-Communist industrialized nations are members, is now promoting a declaration calling for "a new approach to economic growth that will take into account all components of the quality of life, and not only the quantity of goods produced." This is a real departure from the goal of unrestricted growth.

D. U.S. NATURAL RESOURCES, EXHAUSTIBLE AND RENEWABLE

U.S. citizens are suddenly being asked to modify their faith in the bountiful resources of their country. They learned in school how early American explorations revealed incredible riches in forests, farmlands, minerals, and fuels. The wealth seemed to be endless: iron in Michigan and Minnesota, gold in California and the Black Hills, copper in Montana and Arizona, oil in Pennsylvania and Texas, silver in Colorado, lead in Illinois, zinc in Missouri, and lumber and coal almost anywhere they were needed. As exploration and development continued, more oil was found, more copper, and more coal, and then natural gas, aluminum, nickel, molybdenum, borax, and magnesia. As the demands grew, so did the supplies. The same was true of crop resources, not only for food, but also for paper, cordage, and fabrics and textiles.

With the increasing improvement in living standards and growth of industry, ordinary citizens gradually became aware that some needs were being supplied by imports from abroad; but they have failed to notice that in recent years the supplies of more and more U.S. raw materials have necessarily been supplemented by imports. Demand has been outrunning the native supply, and there are some important minerals we do not have at all. Imports of ores, minerals, and chemicals have kept increasing without much public notice; until, in the 1960s, it became necessary to import petroleum and natural gas. By 1972, we were importing over 20% of our petroleum needs, in spite of a U.S. production of 150 billion gallons per year (which is about 100 gallons per month per motor vehicle).

When oil supplies from the Near East became a political weapon, the American public awoke to its dependence on imports for a great many strategic materials. The inadequacy of U.S. supplies actually reflected certain world shortages also, and the entire subject of scarcities is beginning to attract some attention. Energy supplies remain the most critical (Chapter 3), but food is fast becoming a world problem (Chapter 4). Within the United States, the most obvious scarcity of nonrenewables, after fossil fuels, is that of metals. Price increases have for some time indicated shortages. For example, from 1965 to 1973, the increase in the domestic price of zinc was 170%.

Exhaustion of *world* supplies of materials is certain to become a topic of some concern as populations increase. In the past, most shortages have been local, often because of poor distribution of heavy or bulk commodities, and sometimes in relation to trade balances and financing.

Known world resources of metals and minerals must now be carefully appraised and compared with rates of use. A partial comparison is given in Table 2, in which the present U.S. consumption, production, and reserves are compared with those of the rest of the world. This 1960 study indicates that the United States is not in a safe position as regards platinum, since we are responsible for 70% of the world's consumption and our production and reserves are negligible. The platinum and other metals used in catalytic converters for new cars is making this worse; and although world reserves are probably adequate, the United States is in a vulnerable position.

Four other metals for which U.S. consumption is 25% or more and production and reserves less than 5% of that of the world total are chromium, cobalt, nickel, and tin. The case of tungsten is unique, since the United States has only 5% of the world's resources, uses about twice what it produces, and over 70% of the world's reserves are in Communist China.

For several metals, the prospective demand over the next decade or two cannot be met without substantial price increases or some new method of extraction. This is true of aluminum, copper, tin, and zinc. The United States uses about one million tons per year of zinc and about seven million tons of aluminum plus copper. These two metals are partially interchangeable in use, so the quantity of each will vary as the prices seesaw upward; but the total consumption will probably continue to increase. Aside from reduced consumption, there are two obvious methods for extending the life of mineral resources: (1) increasing the price, and (2) reclaiming and recycling.

Increases in the price of a virgin metal have wide effects, many of which are difficult to anticipate. Prospecting for additional sources of ores will soon follow, as will efforts to obtain a higher yield of the metal in smelting and refining. New processes may be developed for the use of large quantities of low-grade ores, for example, the pelletized taconite process on the Lake Superior shore added a vast capacity to supplement the richer iron ores from the Mesabi Range. If the newly high-priced metal is an alloying element for iron, copper, or aluminum, the high price may cause a major shift in the use of several alloying metals, with resultant effects on the prices of both pure metals and alloys. A price increase may result in major substitutions; for example, if the price of copper goes very high, much more aluminum will be used for electrical conductors, tubing, and heat exchangers. Price increases can result in changes in metal usage for spraying, plating, powder metallurgy, bearing compositions, and so on. In fact, there are few fields in which a price increase so quickly arouses the technical ingenuity that provides early adjustment. The market results are almost unpredictable, but any substitution process does cost time and money.

Recycling is an expensive process too. Someone should estimate how much it cost in human effort to collect, transport, and process the 1.6 billion aluminum cans that were recycled in 1973. Of course, recycling always saves heat and power as compared with the production of virgin metal from ore. Large amounts of some metals are already being recycled: for example, iron, copper, lead, and tin; and this practice can and should be increased. Other metals are more difficult to reclaim: for example, zinc and mercury. Still others are used mainly for plating and/or alloying, and recycling them is next to impossible, for example, chromium, cadmium, and vanadium. Another typical example of an end use that does not lend itself to much recycling is the wide use of silver for photographic emulsions.

American mass-production industry is accused of "designing for obsolescence"; and this practice has certainly increased the piles of mixed-metal scrap, especially of automobiles and major appliances. If such consumer goods were instead made more durable, assembled from standard parts, and easily repaired with no special tools, junk piles would be much smaller and more easily recycled. Manufacturers could be required to provide for easy identification and separation of the materials, with trade-in and reclaiming in their own plants. Recovery of nonrenewable materials could eventually replace much of the present consumption-and-waste sequence, provided the necessary changes in attitude could be brought about. But the time and money thus expended would not then be available elsewhere.

What about the effects of population growth? If world population doubles in the next 30 years, and even today many shortages of metals and minerals are already appearing, is there any real alternative to reduced consumption of these strategic materials? The demands will also certainly increase as more and more nations industrialize.

Wider use of renewable, crop-type materials as substitutes for metals is an alternative that must be given more attention. In uses like structures, frames, rods, sheets, and wires, nonmetallics can often be substituted for metals. As population and corresponding industrial production increase and metals become more expensive, wider use of wood, glass, ceramics, cements, concrete, masonry, and the many crop fibers must be developed, taking the place of metals. Synthetic fibers and molded parts and fabrics are not generally acceptable as substitute materials because they are based largely on petrochemicals. This means that wool, cotton, silk, and flax may again come into wider use.

By increasing the output of plant-crop materials such as trees and the many fiber crops, even a growing population can be supplied with these renewable materials. This, however, assumes wise long-time planning and the availability of land and fertilizers for this purpose. But cropland is definitely limited, as indicated in Chapter 4, and must be largely reserved for food production. Food crops have first priority in the land-use program; so that it seems unlikely that, say 25 years hence, greatly increased land areas for fiber crops can be made available. Present crops of wood, tree fibers, and cotton are very large, however (Table 6)[7] and could even be increased on the same land. Note that these crops have increased very little since 1950. Cheap lumber, plywood, paper, and cotton fiber have contributed to expansive and wasteful U.S. use that would shock most other countries. Many shifts in ultimate usage are possible, such as reserving saw logs for high-grade uses like trim and furniture lumber, and substituting masonry and waste materials for much of the building lumber now produced for framing, sheathing, and flooring. There is every reason to expect that with almost a billion acres available, wood and fiber crops can be greatly increased by more intensive methods. Wider planting of fast-growing trees and coarse-fiber

[7] Table 6. U.S. Crops of Woods and Fibers, 1950–1973: Land Use, Production, and Export.

plants can serve the paper, boxboard, pressboard, and cordage industries. Wider use of waste and more recycling would add to the output,
even providing some fuel.

A question could be raised here that deserves careful consideration,
namely, how much time and effort should be diverted to recycling and to
developing substitute materials when the working population is increasing only slowly? Labor thus diverted may aggravate a personnel shortage in other essential fields. If, however, world population growth continues at present rates, no amount of extra effort by more and more
people can indefinitely continue to increase the relative supply of the
metals, minerals, and woods that have helped to make present-day
living standards possible. How large a population is eventually going to
compete for these scarce materials?

As U.S. population increases, what are the prospects of larger
domestic crops of nonfood products, wood, paper, cotton, and tobacco?
Improvement in growing and harvesting methods is the most promising
source of increases. Two other avenues are available, namely, increases
in nonfood crop acreage and larger net imports.

At present, commercial timberland occupies only two-thirds of our
total forest area (Table 6). A small portion of the remaining one-third
might be commercialized. Great areas of native forest were cleared by
early settlers, especially in the northern hardwood regions. There is
some opportunity for reforestation and for extension of present forest
areas to provide more commercial timberland.

Crops of cotton and tobacco occupied over 25 million acres during
1951–1955, and now they occupy only about 15 million acres, of which
roughly a million acres are for tobacco. This downward trend might be
reversed if necessary.

There are prospects for substantial reductions in exports and increases in imports of almost all nonfood crops. This conserves our own
land resources. Consumption of forest products in the United States has
not varied greatly from $3.5 billion a year (1970 dollars) since 1950, and
recently about one-fifth of this total has been imported. Exports are about
one-half as large as imports. Roughly one-third of the cotton crop is
exported, and imports of baled cotton are small. In addition, the exports
of other fibers, wool, and yarn are substantial (Table 6). More could be
retained for domestic use. At least one-third of the tobacco crop is
exported.

As food exports increase, any increase in net imports or reduction in
exports of nonfood crops can be an advantage. Any such shifts must be
taken into account in terms of the U.S. balance of payments, because oil
and food are the two large items in U.S. foreign trade.

Another contribution to the total crop volume under U.S. control is represented by the increased use of foreign land by American operators. Several U.S. corporations already have such projects in operation, and there are additional opportunities. This total is relatively small, however, especially when the products are sold abroad rather than being shipped to the United States.

In spite of the many proposals for conserving our mineral resources by the wider use of substitute materials obtained from agriculture and forestry, such possibilities are in no sense unlimited. Most good, well-watered land is already in use, and marginal land calls for substantial capital investment. Larger crops can be produced on present land, but this takes more labor, capital, water, and fertilizer. In view of the world shortages of fertilizer, especially nitrogen fertilizer produced from natural gas and petroleum, this could easily become the limiting factor. This is another example of the interdependence of our critical problems.

E. MALTHUS AND THE U.S. FARMER

For anyone who knows the great production potential of American farms and who remembers the vast stores of surplus grain on almost every farm, it is most difficult to shift to any thought of shortage. In fact, there is no prospect that a shortage of food in the United States will ever be due to limited productive capacity. High food prices and temporary shortages have been experienced on several occasions, sometimes caused by government farm planning and regulation to increase the income of farmers, or by overexporting. But in many other countries food supply is a chronic problem, and poor crop seasons cause tragic suffering.

In 1798 Thomas Robert Malthus published his famous *Essay on Population*. He revised and elaborated it through six editions to 1826, and this essay has made people somewhat apprehensive ever since. His basic thesis was that population expands in a geometric progression and always tends to outrun the production of food. No doubt Malthus qualified as a pessimist, but his basic idea has been too logical to discard completely. He maintained that population always expands to the limit of subsistence, and that disease, war, and famine are the ultimate population controllers. History affords a long story of attempts to balance the production and consumption of food by storage programs, duties, subsidies, prohibition, penalties, and specific regulation. Malthus was familiar with the early English Corn Laws (dating back to about 1180 and not repealed until 1846), so he was well aware of price fixing by duties and bounties and the resulting troubles. The Malthus essays were theoretical and have been ridiculed in the twentieth century, but their great impression still lingers and finds new application in the recent statistics on malnutrition and famine in Africa and Asia.

The problems of the food-price-population balance persist and have not been solved through elaborate U.S. government programs which started with the Agricultural Adjustment Act of 1933. Designed largely for keeping the U.S. farmer's income balanced on a parity with that of other groups, these programs have never anticipated the great food shortages and malnutrition that now exist in large areas of the world, nor the explosive growth of population that has been responsible. Food shortages have happened before, mainly in Asia; but this time the long-term prospects look frightening.

As with the energy shortage, we have not been without early warnings of impending food scarcities. In the midst of American abundance, all such warnings are difficult to accept; and Congress has wrangled about parity incomes for farmers and subsidies for not growing crops. The profit system is of little use in solving the problems of the underdeveloped countries where there are too many people, too little food, poor agricultural land, and poor transportation. Governments are not much better, because they seldom engage in long-term planning. The habit of the U.S. Congress of spending full time on the emergencies of the day is also the habit of almost all other governments. International attention and the actions of the UN and its Food and Agriculture Organization (FAO) have created some impression; but support of the UN is weak and scattered, and the political demands of the individual member nations always seem to take precedence. What, then, are the long-term prospects for adequate food supplies?

Several books were published during the 1960s that had as their sole purpose warnings about widespread famine that would soon occur in many parts of the world, predicted for as early as 1975. We now realize that these predictions were justified, and we are especially concerned with the prospects for the next 10 to 25 years. We fervently hope that population growth can be abated somewhat, and we want to know whether technical and economic means are available to prevent increasing famine in our lifetime. What measures are necessary to increase the food supply in the critical areas? Will the people involved accept these changes? What are the opportunities for action on our part? Predictions of disaster are hard to accept after so many generations of good times in our own country, but each year there are new statistics of malnutrition. Table 7[8] contrasts the average diet in ten poor countries with that in ten

[8] Table 7.   Food Supplies, Affluence versus Poverty, 1970: Upper and Lower 20% of the Countries of the World.

affluent countries, each group representing about 20% of the world's population. (A similar comparison could be made for over a billion people in the poorer countries, from Table 1.)

The poorer group in Table 7 has a weighted average diet of less than 1950 calories per capita per day, and that diet is also low in protein. Protein is ample in diets containing a high percentage of foods of animal origin (meat, fish, dairy products, eggs), but legumes and nuts also furnish high protein. Daily calorie requirements for good health cannot be stated as a specific number, but a minimum of 2500 calories is sometimes used. In estimating the average of calories required for an entire population, the age distribution should be taken into account. For smaller groups it may be necessary to recognize occupation, climate conditions, and types of food. Specialists in nutrition have made estimates of minimum diets for specific national populations, and these range from 2200 to 2700 calories per day, or higher.

The full mechanization of agriculture and of food acquisition from sea and land is a process that has been accomplished in several areas in our lifetimes, but this is a one-time miracle to which has already been added vast production and use of fertilizers and pesticides and the addition of new varieties of high-yield crops. These have added to the plenty that the advanced nations have enjoyed, but there is no other set of advances on the horizon that will similarly benefit subsequent generations. Much new land has been farmed and distant fishing areas have been occupied, thanks to the tractor and the motorship.

Since the population keeps on growing, the prospects for any new means of increasing the world's food supply become vitally important. Malnutrition and famine are today's warning signals, and it is time to examine the facts of food production worldwide and to determine what can be done to increase and improve the food supply to meet the future needs we can already predict. The United States is in a safe position, but even we have not solved the problem of ample diet for all our people. A Senate report recently stated that 12 million Americans continue to be malnourished. Planning for future food is urgent, and it centers on two major possibilities: improving agriculture in the backward countries, and sharing the abundance of food that can be produced on all possible land in agriculturally advanced countries such as the United States and Canada (see Chapter 4).

With the United States now in such an important position with respect to the world food supply, it is necessary to recognize the vast capabilities of U.S. agriculture and the rapid changes taking place. The combined effects of the disappearance of former "surplus" stocks and the growing demand from other countries have emphasized the necessity for both present control and future planning for food production and exports. Since massive U.S. imports of fuels, metals, and raw materials are necessary, grain exports have a welcome effect on the nation's balance of payments.

When large crop exports represent a substantial part of the farmers' market, the domestic food price structure is disturbed and crop planning for the farmer becomes difficult. The fertilizer situation is already critical and promises to become worse. Major expansion of U.S. crop production would introduce further world shortages of fertilizer. Vast grain exports also take with them important soil nutrients, thus dissipating a valuable natural resource. Expanded farm activity calls for more fuel, too, and aggravates the energy shortage.

A fair picture of the food production situation in the United States can be seen in Table 8.[9] Large-scale mechanized farming has continued to increase, and the average farm size has doubled in the last 30 years. During 1970–1973 the total employment on farms was only about one-half of what it was in the early 1950s. Since employment in food preparation, transportation, and sales has been increasing, there are now many more food handlers than there are food producers in the United States. Nevertheless, the farmer's 40-cent share of the consumer's food dollar has not been reduced (U.S. Department of Agriculture statistics, 1957–1974).

With the land area for pasture and grazing twice that for crops, and about one-half of the crop produced used for animal feed, the high cost of a human diet of meat, dairy, and poultry products is emphasized. The United States also imports a fair quantity of meat, although total agricultural exports vastly exceed imports. Almost one-half of total fish production and imports is unavailable as table food, being used largely for poultry and pet food.

It seems likely that, as world population continues to increase, the United States will be under increasing pressure to reduce the percentage of meat in its diet. Feed production has doubled since 1960. The U.S. contribution to the world food supply could be greatly increased if the nation's population of meat animals and pets was reduced.

A major difficulty arises in the adjustment of food supplies to consumer needs. Grain reserves through carryover storage can be used to reduce wide price fluctuations, as well as to compensate for crop failures. The building of food reserves by all major nations is now being urgently advocated. In U.S. farm production areas incentives are very important. There is good reason to favor continuation and improvement of farm support, parity, and price-control programs, even if subsidies are required. Market forces alone are not adequate to meet the complex

---

[9] Table 8.   U.S. Food Production, Consumption, and Trade, 1950–1973: Acreage, Employment, and Value—Field, Animal, and Fish Products.

demands, and maximum production cannot be expected from an unregulated profit economy for food production.

Ensuring continued incentives for the farmer is not easy under our free-enterprise system, but government controls are resisted and a control program often leads to abuse. Production incentives are especially difficult under conditions of more food than we can use, but this is not a new problem, nor are we the only ones that face it. The smooth flow of food grains to consumers that need them involves many factors entirely beyond the control of the farmers. The complexities of international trade are encountered in purchases and shipments, and with today's volumes they have a major effect on national trade balances.

Financing and distribution of foreign aid plays a large part, since America has for some years been the source of over 75% of all world food relief. Individual farmers certainly have little control over this crop export program and they have no assurance that their products can be moved steadily to market at harvest time. They resent the high wages and the restrictions on productivity in the negotiated contracts of many who transport and handle their products. How can the 4% of the population that are farmers, loosely organized and spread into every corner of every state, even get the attention of the public about their troubles? They find it necessary to demonstrate by destroying calves, milk, pigs, or other products, but this arouses much antagonism. All these difficulties tend to dampen the incentives to take big risks and to exert extreme efforts toward increasing our food crops.

Whether U.S. and Canadian farmers can disprove Malthus depends on the response of the American people to many opportunities, but the ultimate key is the rate of world population growth. Food production in all countries is also critically dependent on the cost of energy. Environmental pollution occurs in all fuel-energy use, and strict measures to reduce this pollution increase energy prices. It is urgent that the interdependence of the crises in food, energy, environment, and population be more widely recognized. Vigorous remedies can then be coordinated according to a systems plan, after which they will hopefully receive the necessary public support. More widespread appreciation of world food needs is a first requisite. Only then will the plans and decisions to help meet these needs receive the attention they deserve. Here is an opportunity for all of us.

F.   POLLUTION CONTROL AND ENVIRONMENTAL MOVEMENTS

These important movements, initiated and carried forward by devoted and single-minded enthusiasts, have gained great momentum and resulted in impressive accomplishments. Recent campaigns for control of pollution and protection of the ecological environment are accomplishing much more than might have been expected. Conservation and better use of natural resources was a cause that excited such early enthusiasts as Theodore Roosevelt and Gifford Pinchot. It is finally being given much more widespread support, thanks to several pollution crises and to shortages and high prices of natural materials. In fact, accomplishments in these fields in the last dozen years have been amazing, and the reasons for this are varied and interesting. How was so much accomplished in this short time? Perhaps the results are small when measured against the total objectives, but the changes demanded in our personal and environmental habits are so extensive that complete success cannot be achieved quickly.

The environmental movement had its beginnings in long-time concern about pure drinking water and the nuisance of smoking chimneys and foul exhausts. The 1960s saw an awakening to the increasing danger of all types of environmental pollution, including noise and ionizing radiation. Gradually this thinking embraced the earlier movements designated "conservation" and "safety" and adopted the broad objectives of protecting humans in all environments and embracing the entire ecology.

Thousands of books and pamphlets on various environmental subjects have been published since 1960, and no less than fifty magazines and serial publications have appeared. The energy crisis accelerated the formation of committees and departments in government, business, education, and voluntary associations, all charged with attending to environmental problems. Many active advocates and semipolitical groups have appeared and have been influential in securing the adoption of legal measures, for example, the postponement of the Alaska oil pipeline and the enactment of laws at federal, state, and local levels. The National Environmental Policy Act of 1969 and the Occupational Health and Safety Act of 1970 were especially significant, because of their wide coverage and the enforcement activities that have followed.

The control of pollution of air, water, and land is very closely related to the supplying of energy and to its uses (Chapter 3). This is illustrated by many of the legal controversies over such subjects as strip mining, oil spills, automobile pollution, thermal pollution by power plants, incineration of refuse, the Alaska pipeline, treatment of high-sulfur fuels and their combustion products, and siting of refineries and nuclear power stations. The arguments have been intensified by suggestions that during the acute energy crisis pollution control should give way to fuel saving.

These controversial struggles illustrate one of the unique and major strengths of the pollution-control movement, namely, the availability of legal remedies. It is very difficult legally to restrain the wasteful use of our natural resources and to make laws to conserve supplies of scarce materials such as minerals, metals, lumber, and paper. The legal prohibition of pollution, however, proved to be a fairly straightforward procedure. Ecology enthusiasts have become well aware of this, and their successes in taking advantage of it have been phenomenal. Not only do we now have strict regulations for the reduction of air pollutants from motor vehicle exhausts, and federal prohibition of pollution of lakes and streams, but state legislatures and city councils have enacted an amazing number of local prohibitions by law and ordinance. Some of these may have involved massive campaigns for cleaning up a city's air or for protecting wildlife from the intrusions of large plants or pipelines; but in any case the end result is legal prohibition or legal specifications for minimum allowable pollution. Communities on the East, West, and Gulf Coasts and in the states of Oregon, California, Colorado, and Florida have been especially aggressive in these respects, with no little assistance from the Sierra Club and others. Several local limitations against town or city growth have been enacted.

One of the great inventions of the ecologists has been the requirement for an "environmental impact statement." This holds up the start of any project that threatens pollution of the environment until a full quantitative study has been made of the threat or possibility of damage to the environment; and this full study is reduced to writing and formally presented with the best possible evidence as to whether the legal limits of pollution, or interference with the environment, will or will not be exceeded. Many complaints have been made about unnecessary delays and the postponement of urgently needed development and construction by environmental impact investigations and similar tactics. It is a fact, however, that in the past great construction projects have been started before adequate planning was completed. More recently, engineers and other technical planners have been overruled and required to complete their homework concerning environmental effects before permits to start projects were issued. In the meantime, anxious businesspeople and promoters see committed funds unused and community needs unmet by reason of these delays.

The energy crisis has precipitated a major conflict between the environmentalists with their legal weapons on the one hand, and the economic community on the other, over the question of how much and how long environmental controls should be relaxed in order to allow greater production of fuels and the continuing use of present fuels with existing equipment. Such conflicts can be expected to continue for some time. The sharpest disagreements have been in connection with automotive fuels and high-sulfur oils and coals. The whole matter is complicated by the inadequacy of convenient native U.S. supplies of low-pollution fuels where and when they are needed.

The 1960s was a decade of environmental accomplishment. A great number of membership clubs and societies, by means of wide collections of dues and contributions, were able to finance intensive educational campaigns and specific lobbying in favor of federal, state, and local enactments for pollution control. The Air Pollution Control Act of 1962 and the Clean Air Act of 1963 (amended in 1965, 1967, and 1970) resulted in programs of investigation and enforcement. The Water Pollution Control Act of 1956 was amended in 1961, 1965, and 1971, and the Water Pollution Control Administration was set up in 1966. It became the Water Quality Administration in 1970. At this time the Water Quality Improvement Act also provided tighter controls. Several other types of pollution control were administered by federal agencies, including the Bureau of Solid Waste Management, the Atomic Energy Commission (AEC), the Federal Radiation Council, and the Pesticides Research Program.

Efforts of scattered agencies in the departments of Commerce, Interior, and Health, Education, and Welfare, and others were joined by means of the National Environmental Policy Act in 1969, and finally by the acts that set up the Environmental Protection Agency (EPA) and the Occupational Health and Safety Administration (OSHA), both of which have wide powers. OSHA operates with respect to health and safety provisions in industry. EPA has a budget in billions, has offices of Air, Water, Solid Waste, Pesticides, and Radiation, and through its many regional offices is in close contact with state and local environmental agencies. It has a large publication program.

National publicity about such conditions as the Santa Barbara oil spill and the "dead sea" aspects of Lake Erie helped to stimulate legislative acts for environmental control. Great numbers of conferences, short courses, and regional and national meetings were organized by federal and local government divisions, universities, trade associations, technical societies, and local ecology groups to spread information about environmental needs and to extend technical education on ways and means for pollution control.

Most large libraries established environmental collections, and many of these now include thousands of books, reports, pamphlets, and magazines. Publishers set up environmental divisions with technical specialists and consulting editors. The most far-reaching moves for the future of the movement have been the new offerings in universities and schools. College degrees designating environmental specialization are now offered, either by departments of environmental studies or associated with such departments as government affairs, management, engineering, social studies, or urban studies.

It is encouraging that in spite of our habit of intense specialization, environmental studies and research have become largely interdisciplinary. Many computer-assisted projects with environmental aspects, such as those of the Club of Rome, have been accomplished by interdisciplinary teams with broad advisory assistance. Their results have received favorable attention from a wide spectrum of interested groups and have been reported to the general public, with much resulting discussion.

A tremendous change in public attitude and in the business atmosphere has come about as compared with the early days when industry letterheads carried pictures of factories with smoking chimneys to denote prosperity. Unfortunately, most U.S. heavy industry, with its smelters, blast furnaces, steel mills, foundries, and forge shops, were built and equipped prior to the intense environmental interest. They are all heavy polluters of both air and water and cannot easily be changed. Communities like Gary, Cleveland, East Pittsburgh, and Anaconda grew up under a pall of smoke and fumes, and changes are expensive, so much so that all the industries' products will necessarily cost more in the future. The same is true for mining operations, power plants, and proposed shale oil and coal gasification plants. Freedom from pollution is not just a matter of wanting better conditions and passing laws. Many years and billions of dollars are required, and all energy-intensive services and heavy goods are going to take a larger share of everybody's budget.

Some ecologists object to the recent emphasis on pollution and its control on the grounds that pollution is a local disturbance which is curable by local remedies, while much more threatening to humankind are the widespread disruptions of nature's stable systems. An example of the latter is the soil erosion that follows deforestation, overgrazing, and conversion of marginal farmland to crops. These abuses by man convert large areas into desert wastelands, a process that is most difficult to reverse and which is an increasing threat to human food and water supplies. The ecologist is also concerned with modifications of climate by these large increases in desert area and by the large-scale release of carbon dioxide and of waste heat from combustion. Another example involves the encroachment of populations on wildlife areas. This endangers many species of animals, birds, and fish. Each plays some unique part in nature's stable system.

Thanks to these ecologists, as well as to population-control enthusiasts, public sentiment in favor of protection of nature's bounties, land, water, and air, has finally been aroused. The appeal of measures for health protection has been even stronger. The rapid growth of organizations of "concerned citizens" has surprised even the groups themselves. On local, state, and national levels there have been hundreds of successful campaigns to organize for environmental protection. The Sierra Club, the Environmental Defense Fund, and the National Resources Defense Council are examples of nationwide efforts. These activities have had political appeal, and legal regulation has soon followed.

Vast environmental studies are now legally required for projects or plans to run a pipeline, build a dam, or dredge a waterway. The final "environmental impact statement" required may involve a dozen volumes of hundreds of pages each. Even more significant are the public hearings and the local meetings that effectively delay a questionable project, discourage its promoters, and may result in stringent regulation or prohibition. Public sentiment in these directions is easy to organize and highly effective.

Efforts toward environmental protection in a dozen directions have now matured into a much broader movement to improve the quality of life. This viewpoint includes economic and social perspectives, as well as environmental and technical ones. The EPA has instituted several measures to promote this concept. In 1972 it sponsored a symposium for this purpose and published the proceedings under the title, *The Quality of Life Concept: A Potential Tool for Decision Makers* (LC No. EP1.2 Q2, 400 pp. 1973). After attempting definitions of "quality of life" the symposium explored viewpoints of the general public, and of its constituents by age, sex, and race. Viewpoints of the various disciplines were aired. As to the attainment of specific goals for the good life, there was a concern about quantification or measurement. How can the results of a program be measured? Examples of this were given and discussed, but the need for further investigation was emphasized. Included in the proceedings volume is a 300-page anthology prepared for the participants of the conference. Among the points emphasized were the interdependence of various efforts toward improving the quality of life, the need for both regional and national approaches, and the importance of education of the public to obtain support for long-range policies and programs.

Another federal agency emphasizing the quality-of-life concept is the Council on Environmental Quality (CEQ). This organization is resident in the executive office of the president. Under both Presidents Nixon and Ford its contributions have been substantial. Of particular interest here are its two annual publications titled, "Environmental Quality" and "The President's Environmental Program." The former is actually the annual report of the council, written by the staff of the council and submitted by the president to Congress. (This report covers several hundred pages and is available from the Superintendent of Documents.)

The 1974 fifth annual report, released in 1975, was notable for a 17-page "highlights" summary, about 800 keyed references, and almost 150 tables and figures. Compared with earlier reports, this issue reflected the broadening viewpoints of the environmental movement, spreading to embrace such topics as land-use planning and control, and wildlife protection. Trends in environmental policy are discussed, outlining not only developments in the United States but also extending to the global environment, the cooperative programs of the UN, and various bilateral arrangements. (See also Chapter 7, Section D.)

The earth's capacity to tolerate the "load" may be considered in terms of the exponential growth of population and thus of waste production and energy release. If one such process has a yearly growth of 5%, for example, a doubling time of 15 years, and the process reaches the environmental danger point after two centuries, what about the following 15 years? It is often overlooked that with continuing yearly growth of 5% the environment must accommodate itself to as great a load in the next 15 years as it absorbed during the previous two centuries. This seems like a very sudden environmental overload, merely because the danger point has been crossed. Exponential growth rates of 2 to 6%, which have been common in the twentieth century and now threaten the environment, could become catastrophic in the twenty-first century. With continued annual growth at a constant annual rate within such ranges as are represented in Table 10, for example, the burden for our great-grandchildren will be staggering, and the environment will not tolerate it (see Chapter 7, Section D). Here are good reasons for the environmental movement.

G. ZPG—THE IMPOSSIBLE SOLUTION

ZPG, the worldwide proposal, would limit the experiences of parenthood and the reproductive sex function of the entire human race. In this respect it is not only extreme, but it calls for a sudden and widespread voluntary change in human behavior—something that has never before been accomplished. True, there have been fads and fashions widely and quickly adopted, but in no sense were these related to essential human needs and drives. There have been forced migrations, crusades, and world wars—not universal, and hardly voluntary.

The implications of ZPG are so complex and far-reaching that the accomplishments are slow and controversial and these programs earn

almost as much opposition as support. Wide recognition of the importance of population control is, however, gradually dawning. Contraception and abortion, the "hot subjects" of planned parenthood and of the
advocates of ZPG, have affected the thinking of the entire world, with no
small results. The intimate nature of the responses required and the
many complexities of time, place, belief, and tradition will continue to
becloud and delay the support of measures for population control. The
limitations of spaceship earth convince us that ZPG must eventually
succeed, and its advocates should be praised and supported; but it
cannot be accomplished in our lifetimes for several very good reasons,
so we had better plan accordingly.

The increase in birthrates following World War II gave rise to a
familiar term, the "Population Explosion"; but this large group has now
reached the age of parenthood. The rate of population growth, in any
case, obviously depends on the age distribution of the existing population, and specifically on the relative number of females of child-bearing
age.

Plans for the future must be based on predictions that seem reasonable and accurate. After fully examining the means for attaining ZPG the
world over, the facts of today may be translated into future possibilities,
and impossibilities.

We must admit that spaceship earth has a finite, limited capacity for
meeting the needs of its human population; many of those limits are now
being approached, so it seems very logical to prescribe ZPG as an
immediate and permanent remedy. But can we disregard all the factors
that make this solution generally impossible? Such requirements as the
immediate abandonment of our well-established philosophy of growth,
and asking all parents to limit their families to three children or less,
contain impossibilities that should be sufficiently obvious. When we
examine the means for accomplishing these objectives very soon, and
everywhere, the impossibility looms even larger, especially for populations in which younger people predominate. How about populations in
which a large family has always meant success, social distinction, and
fulfillment of the highest religious duty? How about populations in which
modern health care has suddenly cut infant mortality after generations of
large families with few adult survivors? How about parents with honest
convictions that abortion or even contraception is little better than mur-
der? How about people living in parts of Australia or Siberia where the
greatest evident need is for *more* people to help in development? How
about the many dictatorships where more soldiers are wanted, and
where more people are needed to settle in newly conquered areas?

Since the net increase in native population is the number of live births less the number of deaths, the growth rate increases if health conditions and health care improve. In fact, statistics show that fertility, in live births per 1000 population, is much more constant than mortality. The net rate of population growth increases as childhood diseases are averted or controlled, even when there is no increase in live births per 1000 population. As medical care in the less developed countries has improved in the twentieth century, the rate of population growth has markedly increased in these countries. World statistics as of 1970 show a large increase in the youthful population in recent years. In fact, the world's "younger generation" (0 to 24 years) was almost 400 million larger in 1970 than in 1960, and in the United States it was 30 million larger in 1970 than in 1950 (Table 9).[10] In the developing countries this group is 60% of the entire population, and worldwide it is 55%. This age group increased by 326 million in the developing nations during the 1960s, while in the developed nations the increase was only 70 million. We hear much about an "aging population," but in the United States the position of the 0 to 24 age group increased from 41.6% in 1950 to 46.0% in 1970. This trend promises to change, but there can be no doubt about a growing U.S. population for the rest of the twentieth century.

There have been many reasons—economic, social, and religious— why large families have been highly regarded. The wonder of, and reverence for, human life has been a strong religious influence. Economically, the large family is a basis for family enterprise and for its continuity, especially on the farm. Socially, the large family is a strong unit and continuously provides important social contacts as the children grow to maturity. High regard for the large family is built into almost every phase of life, and it cannot be suddenly eliminated.

Birth control and abortion will be accepted by some people in some countries, but the means remain controversial. Until birth control methods become simple, sure, safe, easy, and inexpensive, they will be practiced by only a limited number in any given population. Making safe abortion readily available is a long-term project, even in countries where health services are well organized; but a billion people live in countries where there is an average of one medical doctor for almost 1000 population. Moreover, most doctors live and practice only in towns and cities; and they are not readily available even there.

As of the mid-1970s there are many countries in which 80% or more of the population are in the age brackets below forty-five, and only a few countries in which people under forty-five constitute less than 70% of the

[10] Table 9. Youthful World Populations, 1950–1970: Populations Less Than 25 Years of Age (millions).

population. Thus, for at least the next 40 years, the number of women of child-bearing age will be large; and world population will grow. But how much?

Many efforts have been made by national census bureaus, by the UN, and by demographic researchers to answer this question. Since the uncertainties are so great, more than one set of assumptions is usually made for a given study. Both the UN and the U.S. Census Bureau have used the "three guesses" approach, that is, high, medium, and low birthrates, and published extensive tables (see summaries, Tables 3 and 4). Several studies have been made with four or even five sets of assumptions.[11] In general, the conclusions from these careful analyses are that, while conditions are different in each country, there is no possibility that world ZPG can be attained soon.

As to growth in the United States by the year 2000, the Census Bureau projections indicate that the likely population will be between 250 and 300 million at that date (Table 4). These extremes are based, respectively, on averages of 1.8 and 2.8 live births per prospective mother. Since many prospective mothers are childless or have only one offspring, families of three to five will not be unusual for these averages of 1.8 and 2.8, and some larger families can even be expected. Past U.S. experience shows a considerable fluctuation in the population growth, but some of this is due to immigration. It seems very unlikely that an overall growth rate of 0.5% will be attained during 1990–2000, as required to keep the U.S. population in the year 2000 within the limit of 250 million.

A favorite assumption for estimating purposes is a "net reproduction rate" of unity. This assumes that an average of 1.0 daughters survive. In each family the count is made when the daughter reaches the age of her mother when that daughter was born. It is estimated that the worldwide net reproduction rate in 1970 was just under 2.0, and that in the United States it was just over 1.0. In the developed countries the average family size in 1970 was two to three, and in the developing countries five to seven. The net reproduction rate for a given country is not its growth rate, since the latter depends also on the specific mortality curve and on immigration and emigration; but a worldwide net reproduction rate of 1.0 would be a close approximation to ZPG.

The attitude of the people in a country really determines how soon a reproduction rate of unity or less can be reached, and whether the low

---

[11] For an extended discussion see Thomas Frejka, *The Future of Population Growth, Alternative Paths to Equilibrium*, Wiley-Interscience, New York, 1972. Gives tables and calculation procedures.

rate can be maintained. The UN World Population Conference in 1974 very well illustrated that much of the underdeveloped world and even some of the developed nations actively oppose ZPG. This stand will not be changed except very slowly. At the 1974 conference the opposing nations actually represented a substantial percentage of the world's population. There have recently been several articles and books opposing ZPG.[12] Many reasons for attitudes of opposition to ZPG were emphatically expressed at the UN conference; but whatever the reasons, this effective polling of world attitudes is one of the most reliable indicators that world ZPG will be impossible in our lifetimes.

Many serious studies are still being made of the prospects for attaining ZPG, and of the effects to be expected with and without population growth. The conclusions are usually very similar, namely, that growth at present rates will eventually lead to near disaster in terms of supplies of food and materials and of the pollution of the environment, but that the built-in inertia of population increase will make attainment of ZPG impossible within the next 50 years. Many of the extremists, either for or against ZPG, do not take this inertia into account, nor do they admit that ecological balances are in danger. The failure to shift from a labor-intensive economy to one that is capital- and energy-intensive certainly dooms a people to low standards of living, but this fact is also widely disregarded. Short-term economic gains are still possible for certain groups as a result of large families. The educational effort required to change this viewpoint is great, but therein also lies a great challenge and opportunity today.

[12] See Wilfred Beckerman, *In Defence of Economic Growth,* Jonathan Cope, London, 1974. (Almost 300 pp.)

Chapter Three

Energy Technology
and
Energy Supplies

A. MEETING A CONTINUING ENERGY CRISIS

Is there, or was there, a real energy crisis? If a popular vote could determine this, it is entirely possible that a poll of the entire U.S. population would decide: "No, there was no energy shortage; and the fuel crisis was largely contrived by the big oil companies, with the fortunate assistance of the politically minded Arabs."

Actually, we were not without ample warning of a coming energy crisis. During many years prior to 1973 these warnings were voiced by the technical community, by geologists, and by government agencies; and preparations were urged for a possible crisis. They were met by a public attitude of: "We've heard that before."

Fortunately the shock of the oil embargo was sufficient to arouse American leaders in both business and government, and during the two following years many programs of information and action were initiated. Impartial groups such as the National Academies of Science and Engineering and large nonprofit foundations supported new studies of our energy future, and several of their reports are quoted and referenced herein. Encouraging evidence of success in the organization of this effort within the federal government is the development of active cooperation among the several agencies concerned with energy supply and regulation. In mid-1975 a comprehensive 600-page report entitled "Energy

Alternatives," was issued under the joint sponsorship of the CEQ, the Energy Research and Development Administration (ERDA), the EPA, the FEA, Federal Power Commission (FPC), Department of the Interior (DOI), and the National Science Foundation (NSF).[13] Although this report is largely a complex technical and economic analysis rather than a projected future program, its appearance under this joint sponsorship is almost startling, but very encouraging.

This one of our four crisis problems, the conversion and use of energy at increasing rates, uniquely affects all the other three crisis areas. Most air and water pollution, for example, comes from our great power plants, from the energy industries that produce steel, copper, aluminum, fibrous products, plastics, chemicals, and synthetics, and from the millions of combustion engines we use to drive everything. Noise and radiation pollution come from the same sources, and most solid waste is made up of energy-produced materials such as those just mentioned.

The material in this book, specifically the present chapter, is intended as a concise summary of the present status and future prospects of our energy supply. Within the limits of about 60 pages and a few tables, both present and future energy sources are considered in quantitative terms.

In the underdeveloped countries, food crises follow rapid population growth because of the low productivity of an agriculture that depends so largely on human and animal energy. Starvation in many of these countries is being warded off largely by food from the United States, Canada, Australia, and New Zealand. But the great amounts of energy we use in producing and transporting this food aggravate our air, water, and land pollution. Millions of tractors, combines, trucks, trains, and ships must be supplied with fuel. But large-scale farming and miracle yields would not be possible without fertilizer (from petroleum), pesticides, machinery for planting, harvesting, and crop drying, and great storage elevators. All these contribute to pollution, but so do the factories that furnish steel, machinery, fertilizers, and pesticides. The more the energy crisis is examined, the more interdependencies appear, and the greater the conviction that it will last a long time.

Continuance of the energy crisis seems remote to the person on the street as long as there is plenty of gasoline and heating oil. We forget that almost one-half of this supply comes largely from abroad, and could be shut off on signal from the Oil Producing and Exporting Countries (OPEC). Moreover, our energy supply *must grow,* not only to supply a few

[13] "Energy Alternatives: A Comprehensive Analysis;" prepared for the CEQ, ERDA, EPA, FEA, FPC, DOI, and NSF, by the Science and Public Policy Program, University of Oklahoma, Norman, Oklahoma. Government Printing Office, May 1975.

million new workers every year, but also because our agriculture must grow, our industry must grow (if slowly), and our known reserves of petroleum and natural gas are dwindling.

The conclusion must be that real shortages are coming, and that there is no quick remedy. Here we have a big educational assignment and opportunity. The facts must be made widely known, so that appropriate measures can be devised and remedial action taken with wide support and participation. Studies and reports on the subject are plentiful, but skepticism remains. Honest disagreements greatly delay the adoption of effective measures. The energy bill passed in late 1975 was signed with reluctance by President Ford. He had already been convinced by his advisers that energy shortages are much more threatening than the new majority in Congress thinks they are. But the compromises represented by this energy bill are only a start and, hopefully, more effective measures can be expected.

One large opportunity for improvement is in our practices in energy usage. Fuels have been so inexpensive in the past years, especially in the United States, that a great many wasteful uses have been tolerated. In many cases it did not pay to do any better. Total fuel consumption has increased much more rapidly in the United States than the 1% annual growth of the population; in fact, fuel usage grew at about 4% per year during 1960–1973.[14] Since this rate of growth calls for doubling the fuel supply in less than 20 years, it is no wonder that a fuel crisis developed. In the largest energy industry, electric power generation, which had been using roughly 60% of all coal and about one-sixth of all natural gas (Table 11),[15] the yearly increase was almost 8%. This rate calls for doubling the generating capacity and the fuel used every 9 years (see further discussion in Section B).

With the Arab oil embargo and price increase, every country began to realize the necessity for conservation and the limiting of fuel usage. It was reported that several European nations, already using much smaller amounts of oil per capita than the United States, were able to reduce their oil consumption 15% or more during the early crisis period. This was three times the reduction attained in the United States. Our fuel usage is divided rather evenly among the four consumption sectors, with sizable yearly increases in each (Tables 10 and 11). Meeting the fuel crisis thus becomes an obligation for each group in the population.

It would serve no good purpose here to attempt to analyze the

[14] Table 10. Some Annual Growth Rates in the U.S., 1950–1973: Population, GNP, Personal Income, Energy.
[15] Table 11. Distribution of U.S. Fuel Use, 1965–1973. Percentages Supplied to Consuming Sectors.

complex demand-supply relationships for the several energy sources. It should be admitted at once, however, that higher retail prices soon increase the supply by bringing back marginal producers, encouraging exploration for new sources, and providing for higher costs of production and distribution, including those imposed by antipollution and safety rules. Higher prices also encourage research and development in all phases from mine or well to customer and, in addition, provide some funding for the building of improved facilities for production, refining, and distribution. Higher fuel prices affect the entire economy however, and are vigorously resisted by the consumer. All energy services and energy-intensive products soon demand a higher percentage of the budget of every individual, business, and government department. This means a lower percentage for something else; and that is a very difficult adjustment, calling for changes in living patterns and in the practices of many business and government operations.

The United States is in the special and very fortunate position of already being a large producer of oil, natural gas, and coal, and of having immense reserves of coal, lignite, oil shale, and nuclear power materials, plus modest amounts of other energy resources. These assets, however, do not preclude a worsening crisis for the next 10 years, even though technical capabilities are available. There is little indication as yet of widespread conviction that major efforts are necessary. The public has not been impressed by the recent downward revisions of earlier estimates of U.S. petroleum resources. Here we encounter again the philosophy of unhampered quantity growth. The ease of obtaining more OPEC imports quite obscures the urgency of an orderly conversion to other energy sources, even though domestic oil production is decreasing and is likely to drop rapidly in the 1980s.

Efforts to increase energy supplies from coal, lignite, nuclear reactors, and oil shale are gaining very slowly, amidst much bickering about possible environmental dangers. The depression should have lessened the energy needs for essential purposes. Even though a negative growth of the GNP was a reality for the 2 years 1974–1975, there was no corresponding drop in the energy demand.

Adoption of a firm energy program for the next 10 years has suffered much delay, and even the energy industries themselves seldom act in contradiction to the prevailing expectation for a return to the boom times of the 1963–1973 decade. It should now be admitted that the problem is so all-embracing that government participation and stimulation are a necessity. Congressional committees have made several studies, and our presidents have issued various reports and made recommendations to Congress. Within government organizations, the Interior Department, the Commerce Department, the National Aeronautics and Space Administration (NASA), the AEC, and the National Academies of Science and Engineering have each made comprehensive reports. These reports have been widely used and are referenced in various sections in this book. The several energy sources will be examined in turn in the following sections, with statements and conclusions about recommended programs for the next 10 to 25 years.

In spite of an apparent easing of the energy crisis during 1974–1976 by massive imports, the United States is bound to experience increasing energy difficulties for many more years to come, probably until 1990 or later. The reasons for this are both internal and international. Internal fuel supplies are enlarged somewhat by price-and-market forces, but this growth is impeded by suspicion and the clamor about corporation profits, and by the resulting government regulations. In the meantime, a very *slow* start is being made on promising future sources, nuclear and otherwise. Most of these, including oil from coal and shale, and solar and geothermal energy, call for great captial investment, encouraged and supported by government, and also require some relaxation of extreme environmental regulations. Without decisive popular support, business and government actions are held up. Supply sources that require 5 to 10 years for design, development, and construction attain mere pilot-plant operation in that time because of lack of support and authorization. There are some advantages to this. As in the early case of the Alaska pipeline, delayed starts mean better technical design and improved protection against serious environmental damage. The public is better protected in the long run, at the cost of a certain amount of suffering in the interim. The bill for massive importation of oil ($24 billion in 1974) must eventually be met by sending abroad many of the products we need ourselves.

The international energy crisis is much more dangerous. Nations such as Japan, Denmark, and Italy, dependent almost entirely on Near East oil, have little or no alternative. In the meantime, the oil cartel (OPEC) is upsetting world trade and finance, fueling runaway inflation, and driving one country after another to desperation. The OPEC profit surplus for 1974, if continued for years, cannot be absorbed by the ordinary financial and investment machinery; and meanwhile the developing countries sink into stagnation and starvation. Their "green revolution" is canceled by high prices for farm power and fertilizer; any crop failures are a disaster, and their industrial development becomes impossible, yet their population growth continues. Perhaps this is the aftermath of the energy crisis rather than the crisis itself, but no country is unaffected, and strong corrective measures must someday be applied.

The five largest importers of Near East oil are the United States, Japan, West Germany, the United Kingdom, and Italy. In 1974 U.S. imports were about three times those of the United Kingdom or Italy. All other countries' importations combined, except those just named, did not equal the U.S. total. Prospective supplies from Alaska and the North Sea will make only a small dent in the total import needs, although they will change the distribution. New sources in China and elsewhere may eventually have an effect.

The actual magnitude of OPEC income is even difficult to comprehend. It represents the largest peacetime transfer of wealth in modern times. OPEC sales in 1974 were reported as $112 billion. The U.S. bill for foreign oil that year was $24 billion. The London *Economist* noted that all the stock of Exxon, the largest U.S. corporation, could be bought with less than 3 months of OPEC income, and that it would take less than 10 years at the 1974 rate to buy out all the companies listed on the New York Stock Exchange.

The future of the OPEC operations certainly cannot be predicted. It is already evident that, in spite of their income from massive bank deposits and security purchases, the world financial and trade system has withstood the shock much better than was expected in 1973. During the business recession the demand for OPEC oil decreased, and their lavish domestic imports reduced their surplus for investment. Their collective investment surplus has been reported as $59.3 billion for 1974, $41.7 billion for 1975, and an estimated $44.9 billion for 1976.

Some economists insist that OPEC capital should be welcomed, because so much capital will be needed for carrying out U.S. plans for energy, environmental, and food production programs (see Chapter 7, Section B). Others remind us that high oil prices encourage additions to U.S. domestic production and refining plants. One segment of the spending program of OPEC nations could be considered a real threat to future world peace, namely, vast outlays for military equipment. The sad fact is that the United States is the major supplier.

As of 1976 the prospects of early reduction in U.S. oil imports are dim. Most authorities are predicting that they are more likely to increase, possibly even reaching 50% of U.S. consumption. Whatever the trends, and in spite of the measures taken, the United States will continue to be highly dependent on foreign oil for many years and will be in a most vulnerable position.

The nature of the continuing energy crisis in each country can be modified by legal mandates and government controls. Even joint action of the industrial nations is a possibility. World recession temporarily delayed larger total energy demands, and shortages have been less acute than predicted, but this does not cancel the likelihood of great difficulties in the future.

B. FUTURE ENERGY DEMANDS

World use of energy has been increasing very rapidly; and if some of the developing nations should attain unexpected success in industrialization, there is no way of predicting the demand explosion that would result. Energy demand is closely proportional to national product output, not only in manufacturing and construction but also in farming, fishing, food processing, and goods distribution. Thus every economic success in a developing country increases its reliance on fuel energy; and if its native fuel resources are inadequate, potential instability is established. Dependable fuel supplies and trade access thereto are essentials for world stability, and they are everyone's concern.

Almost 95% of the human family lives outside the United States, yet our use of energy (fuels) approaches one-half that of the rest of the world combined. In terms of proven fuel reserves, our resources are modest. The rest of the world has known petroleum reserves more than twelve times as great as ours, and probably six times as much natural gas. The United States has plenty of coal, estimated to last 500 years or more. Massive coal deposits are found elsewhere too, but most of them are in the Communist countries.

If it were possible to increase U.S. oil and gas production to the level of present domestic consumption, widely quoted statistics about our known underground reserves of these fuels indicate that these reserves would be exhausted in the 1990s or soon thereafter. There is no danger of this. Production cannot be increased that much; and no doubt more oil and gas will be discovered, especially on the continental shelf off the American coasts.

Energy trends and policies in this country have an almost controlling effect on the world energy market. If the United States actually practices energy conservation and works toward self-sufficiency, and world peace is maintained, the large fuel supplies from the Near East and elsewhere will be ample to meet the increasing demands from the rest of the world. Violent fluctuations in local energy prices may occur, since the supply-demand-price relationship responds to short-time changes; but the long-range prospects are for inherent stability and fair prices, except when energy supplies are used as a political weapon.

Prior to 1970 many projections for the future use of energy were based on simple extrapolations of the 1950–1970 curves for energy use. As shown in Table 10, the actual yearly growth rates of energy consumption in the United States rose into the range of 3 to 10% per year during that period. It is now accepted that the continuance of such rates of growth, which may call for fourfold increases in much of the energy production and plant before the year 2000, is quite impracticable. Perhaps with some success in reducing the rate of population growth, there is reasonable expectation that modest energy growth rates over this period can be successfully met, and that the necessary changes in the habits of energy users will be acceptable. In the following sections each of the available energy sources—petroleum, coal, solar energy, nuclear reactors, and so on—are examined separately with respect to resources, growth, consumer demand, and national self-sufficiency, with some additional comments about the world situation.

As the United States looks forward to lesser dependence on imported fuels, its planning for domestic production must be based on reliable predictions. Energy demands have been increasing more rapidly than the GNP, but the two are interdependent and future energy planning should be based on extrapolations of both. A great number of such extrapolations has been made, and some will be quoted; but the growth rates shown in Table 10 should be kept in view in making any interpretations, and it should be kept in mind that recent rates of growth have been possible only because of large importations. The energy crisis, the ZPG movement, and the business recession have changed the highly optimistic predictions of earlier years.

A comprehensive study of probable energy trends to the year 2040 was reported to the NSF in 1971. This 1500-page report reviewed all published projections prior to that date. After studying fifty-six separate forecast estimates, it predicted a uniform annual growth rate of 2.81% per year for total U.S. energy usage, giving an average expected doubling every 25 years. A steady growth of 5.46% per year was selected for the electric energy supply, doubling every 13 years.[16]

In December 1973 the chairman of the AEC submitted a report to the president, at his request, entitled "The Nation's Energy Future." This report was a condensation and summary of two broad-based studies,

[16] *The U.S. Energy Problem,* 3 vol., 1511 pp., submitted under NSF Contract No. C-645 by Inter Technology Corporation and distributed in 1972 by NTIS (National Technical Information Service, Springfield, Virginia).

one originating from a group of energy workshops sponsored by Cornell University, and the other resulting from sixteen technical review panels consisting of members drawn from thirty-six federal agencies, each panel assisted by about ten consultants from the private sector. They based recommendations on a "projected annual growth in energy consumption of 4.1% from 1972 to 1980, and 3.9% from 1980 to 1985."[17]

Another forecast of future energy growth is summarized in Table 18 (see Section F). Here the projections are based on an average annual increase in energy supplied from all sources of about 3%, calling for a supply in the year 2000 of about two and one-half times that in 1970. It is anticipated that domestic supplies of both petroleum and natural gas will shrink considerably. Petroleum consumption is expected to increase steadily at about 2.2% per year, with supplemental sources filling 50% of the demand at the end of the century, at which time nuclear power is expected to furnish about one-fourth of the total energy. No new energy sources are included in this projection.

The fastest growing segment of the U.S. energy industry has been that of electric power generation, which grew at a rate of 7% or more throughout the period 1950–1973 (Table 10). Even with a considerable reduction in these rates of growth, it appears that in the future electrical utilities will need much more capital per year than the total rate of investment in recent years for all manufacturing industries combined.[18] (See further discussion of capital requirements in Chapter 7, Section A.)

Forecasts of growth uniformly predict that the electric power industry will continue to grow more rapidly than other segments of the energy industry. The Office of Economics of the FPC predicted in 1972 that the energy input for electricity generation in the year 2000 would be more than four times the consumption for that purpose in 1975, or an annual increase of about 5.9%. A like estimate was published by the General Electric Company in 1974. More recently, many respected authorities who readily admit that the overall rate of energy consumption must be reduced in the future, have advocated a program of even greater dependence on electric power. They give many reasons why the doubling of electrical generating capacity at least every 10 years should continue (about 7% annually). This expansion of electric power is required for electric automobiles, for all-electric homes and commercial buildings, for the many electrical processes in industry, and for electric trains and

---

[17] "The Nation's Energy Future," a 170-page report submitted to President Nixon by Dr. Dixy Lee Ray, Chairman of the AEC December 1, 1973. (Superintendent of Documents, No. 5210–00363).

[18] See "Why Business Ran Out of Capacity," *Fortune,* **XC** (N5), 262 (May 1974).

mass transit. There are, nevertheless, major obstacles, such as the siting of plants, their water supply, and the capital and manpower required.

The best use for nuclear energy is in the generation of electricity, and the same can be said for coal. These two primary energy sources are the only ones in adequate supply, and for which power-generating technologies already exist. Highly efficient large plants can be built and equipped so that pollution of the environment is minimal.

Electrical utility generating capacity in the United States in 1970 was 341,000 megawatts (Table 14). If this must continue to double every 10 years (Table 14), the capacity that must be added in the 1970s will be another 341,000 megawatts. The largest plants today are not much over 1000 megawatts. Additions in the 1980s should be equivalent to 682 such 1000-megawatt plants, and in the 1990s 1364 plants. This total, equivalent to 2387 plants of 1000-megawatt capacity each, would be an average of almost fifty plants added in each of the fifty states. The magnitude of this assignment can hardly be appreciated.

The direct cooling water required for each 1000-megawatt, fuel-fired, steam-electric station is of the order of a million gallons per minute at rated load, because the electrical output is not much over one-third of the heat supplied by the fuel. The rest must be rejected to the environment. The heat rejected from a nuclear plant is even greater than that from a fuel-fired plant. Nor could this rejected heat be easily reclaimed, because of the solitary location of most nuclear plants.

This large water consumption calls for plant location on a river, lake, or ocean. For a river, assume that 10% of the total flow could be diverted through the power plant. There are less than thirty rivers in the United States that have the required average flow of 10 million gallons per minute, and there are 2387 plants to be located. Among the rivers that are too small are the Hudson, the Delaware, and the Allegheny. Air cooling by cooling towers is a possible substitue, but their effect on the atmosphere and climate must be considered, and dry cooling towers of this size have never been built.

Recent published estimates of the capital investment required for an electrical utility growth program of this magnitude, that is, 2387 plants of 1000-megawatt size in 30 years, or eighty plants per year, approximate a $2 trillion total. This is about $67 billion per year, and over $800 million per plant (1975 dollars). This figure is very much higher than those reported in the past (see Tables 17 and 35). Inflation must be taken into account in any comparisons, however, and it should be noted that the newer estimates assume a high percentage of expensive nuclear plants and also large investments for environmental protection. Even if this investment estimate averaging $67 billion per year for 30 years proves much too high, it appears that electrical utility expansion at this rate would require a subsidy.

Utility expansion programs call for highly technical personnel, both in construction and operating phases. This is in addition to all the other related demands for technical manpower, such as for greater coal production, for development of new oil and gas fields, for research and development in new energy sources—nuclear, solar, geothermal—and for the establishing of a large new industry for the manufacture of synthetic oil and gas. The needs of the energy industry alone would make it necessary to double the technical and scientific education system. There is little prospect that U.S. engineering and technology schools can meet these simultaneous demands in addition to all those expected from other quarters (see Chapter 7, Section C).

There is every indication that doubling the electrical energy production every 10 years for the balance of the century is an impossible goal. In this book the preferred assumption selected for electrical utility growth, for 1970–2000, is an average of 4.5% per year, or a doubling of capacity about every 15 years (Table 19). This calls for building only 1023 plants of the 1000-megawatt size by the year 2000, instead of 2387 such plants. (See also Table 37.)

With respect to *world* energy requirements for the near future, the Energy Committee of the OECD made predictions that should receive some attention, because the countries represented in this organization account for nearly two-thirds of the *world's* total use of energy. Their detailed predictions of world energy use to 1980 were published in 1973.[19] This report was actually written in 1972, before the Arab oil embargo and the economic difficulties that followed. It anticipated a growth of the "world's primary requirements" averaging 5.6% per year in the 1970s, well above the 4.9% average of the 1960s. This is a rather surprising prediction. If this rate continued to 1982, the total supply that year would be twice what it was in 1970. The committee specifically anticipated "accelerating requirements in the developing countries." The economic recession, and especially the quadrupling of OPEC oil prices, destroyed this prospect. The OECD report also anticipated that oil and gas would furnish a much greater percentage, and coal a much smaller percentage, of total world energy than in 1960.

Considering the long-term future of the petroleum supply, it is probably a good thing that the 20-year growth from 1960 to 1980 cannot possibly result in a tripling of the 1960 usage as predicted. However, this reprieve has been at the expense of privation for the poorer countries.

Returning to the topic of future energy demands in the United States,

[19] "Oil, the Present Situation and Future Prospects," OECD, Paris, 1973.

this author confidently expects that slower growth rates will of necessity prevail after the economic recession. Our projections are given in Table 19. The usual simplistic assumption of a constant growth rate has been used, that is, the average rate during each of the three decades will be the same. Accompanying an average growth rate for the GNP of 2.5% per year, the selected assumptions as regards energy are: 2% annual growth for total petroleum and natural gas, but 2.5% growth for automotive fuels; total energy input expected to increase at a 2.5% rate, and both coal usage and electric power generation to increase at a 4.5% annual rate.

Even these lower rates will call for oil and gas supplies in the year 2000 to be about 81% larger than in 1970, total energy input 143% larger, and coal supply and electric power 270% larger. (See further discussion of these projections in Chapter 3, Section F.)

C. ENERGY SUPPLIES FROM PRESENT SOURCES

During the early 1970s about 90% of all the energy used in the United States was obtained from fuels; the balance came from water power, nuclear fission, and minor sources. If it is assumed that our objective for the next 25 years is not only to increase the total available fuel supply at a cumulative rate of at least 2.5% per year (Table 19) but also to attain self-sufficiency in fuel supplies well before the end of the century, there is much catching up to be done. Moreover, U.S. fuel demands are very exacting as to the amount of each fuel required; and there are rigid specifications for each grade. For coal, the sulfur content has become very important in the control of air pollution by sulfur oxides; and low ash and moisture are desirable. We burn very little crude oil, and the specific demand for the refined products calls for at least two or three grades each of gasoline, diesel fuel, jet engine fuel, light fuel oil, and heavy (viscous) fuel oil, plus bituminous residues and petroleum coke. The refinery not only furnishes ten or more different fuel products, each to close specification, but the output of each product must match a continually changing demand. Every crude oil input is different. Oils from coal or shale hardly even resemble petroleum. These are some of the oil-fuel supply problems, not insoluble, but difficult.

The overall quantities of coal, petroleum, and natural gas, the rates of consumption, for 1965–1974, are indicated by the data in Table 11. The net importations of fuels are given in Table 12.[20] Prior to the Arab oil

[20] Table 12. U.S. Energy Consumption and Imports, 1940–1972: Total and Per Capita Use and Distribution.

embargo it was often loosely predicted that well before the year 2000 the United States would be importing over 50% of its yearly needs of petroleum. The embargo suddenly prompted a realization that such great reliance on imports would not only be a threat to the continuity of domestic and military oil supplies, but that it would eventually be impossible to make the large payments involved. The price increases of foreign oil have emphasized the latter point. The question is: How can massive imports be avoided when domestic supplies are inadequate, even for current use? Obviously, a simple answer is: Lower the demand and increase the domestic output.

The acute experiences of 1973–1975 demonstrated that modest lowering of demand is possible, but a truckers' strike and other evidences of anger and distrust also demonstrated that the sudden imposition of extreme measures is not accepted without internal dissension. For energy conservation the 55-mile speed limit and the 65-degree home thermostat setting were slowly accepted by some fraction of U.S. citizens, but real effectiveness demands a much more comprehensive approach, especially since environmental problems are so intimately involved. Congress set up guidelines and imposed targets for the automotive industry, but the lead time between a proposal and its final accomplishment proved to be nearly a decade. In the meantime, anyone who seriously considered the dangers of an energy shortage soon became convinced that drastic and all-embracing energy conservation was urgent.[21]

The energy used in buildings has been about one-third of total energy usage, and large savings in this sector seemed promising. Parallel studies by firms in the building industry, engineering societies, building owners and managers, universities, and government agencies resulted in a great many convincing demonstrations of savings of 25% or more in specific installations, when intensive conservation practices were instituted. Conservation programs finally have obtained a good start, but consistent and continuing efforts, well beyond 1980, will be required to even approach the goal of reducing energy consumption by one-third or more. (See further discussion in Section G.)

1. Petroleum Fuels

Private automobiles, two or more to a family, give the ordinary U.S. citizen advantages of mobility that are unheard of for nine-tenths of humanity. We hear repeated assertions that our overabundant supply of

---

[21] See L. D. Conta, "Conservation—The Only Way Out," *Professional Engineer*, **46** (2) 40 (February 1976).

energy makes possible the high material standards of life in America. It might be interesting to examine the assertion that most of this progress can be traced to the combustion of oil. Just a brief listing of the ways in which our lives differ from those of our grandfathers forces us to recognize that the burning of oil furnishes the energy we depend on for most of these changes.

The motorbus and the jet plane add to our freedom to go anywhere, anytime, in air-conditioned comfort, thanks to the oil supply. Railway trains, motor trucks, ships, small boats, campers, snowmobiles, and motorbikes all furnish oil-burning transport. Construction machinery of all kinds, hoists and lifts, outdoor maintenance equipment, down to the lawn mower and the snowplow, all depend on oil-combustion power. It is no wonder that people get excited about an interruption in the supply of gasoline and oil fuels, especially when they recognize that in many sections of the United States most of the fuel for winter home heating, and most of the fuel for electric power generation, is oil. (The Northeast and the Southwest are the most dependent on oil.) As an aside, it seldom occurs to us that we wear clothes, walk on carpets, sit on seats, and sleep on mattresses for which the fibers, fabrics, and foams come from oil as the basic raw material.

When we are told that our proven U.S. future reserves of petroleum are only about twenty times the total amount we consume in a single year, this great dependence on oil for transportation, heat, and power seems a bit precarious. It is easy to start an argument, however, about the extent of our petroleum resources and how long they will last. Even before 1920 a decided scare was caused in the technical community when the U.S. Geological Survey reported a prediction of only a 20- or 30-year supply of oil available. Now our annual consumption rates are about ten times what they were in 1918–1920, and the short supply ahead of us has become a very old story. For more than 40 years the discovery of massive new fields, improved recovery and refining, and imports from neighbors and distant lands have kept up with increasing U.S. demands. The public has come to expect this to continue. They are not even impressed by the statement that the total recoverable oil now expected from the Prudhoe Bay field in Alaska is equivalent to only about 2 years' supply at our increasing rates of use. The great gamble of offshore oil may yet provide for all the needs of present generations of Americans. If so, there are other things to worry about. A modern refinery costs a quarter billion dollars and takes many years to build, and every community says, "Don't build it here." Offshore oil spills have already resulted in many prohibitive regulations. There is no assurance that the new oil sources will provide low-sulfur crude, and the sulfur oxide pollution problem is still very serious. At what price to the consumer can all the new developments and the required protective measures be financed? The costs of motor vehicle antipollution will soon be measured in billions of gallons of fuel consumed by engines of lowered efficiency.

Higher petroleum prices are unsettling the economic systems the world over. Americans will not peaceably accept a European scale of prices until they are convinced that it is strictly necessary in order to finance the required expansion of the domestic oil industry, and it will be most difficult to allay present suspicions about politics and profits in the industry. Nevertheless, the prospects of adequate supplies from U.S. sources are encouraging. In contrast, in many industrial countries that depend entirely on imported oil (e.g., Japan) rising oil prices start an inflation spiral which becomes a near disaster. But the greatest hardships and the most insoluble difficulties in the wake of increasing oil prices arise in the underdeveloped and developing nations. The brave plans for assistance and cooperation from the developed world, through the UN, the Agency for International Development (AID), the World Bank, and the many bilateral programs, are completely upset when poor nations must abandon projects of industrialization and development in order to finance their fuel bills. The OPEC countries are getting all the money, and it is said that by 1980 their income from investments in the developed world will equal their income from oil exports.

Of the world's known petroleum reserves, the United States is usually estimated to have 5 or 6% (Table 13), which is just about our proportion of the world's population. Since we are accustomed to consuming more nearly one-third of the world's annual supply, it is very evidently urgent that this imbalance be corrected; or the United States will run out of oil while there is still ample supply elsewhere. We are now convinced that domestic oil is much preferable to increased imports, hence we should promote and expand our four alternatives: (1) petroleum exploration; (2) oil from substitute sources, coal and shale oil; (3) improvements in extraction and refining; and (4) higher efficiency of utilization. In addition, we must actively promote the development and use of other energy sources, especially nuclear energy and solar energy.

Non-Communist Europe, with a population more than twice that of the United States has less than one-half the oil reserves that we do. Through taxation and resulting high prices, they have largely held down consumption (and oil imports). They are now becoming nervous about the Communist nations next door who, for a lesser population, have six or eight times as much oil (see Table 13). The great Asian nations, with one-half of the entire world population had almost no petroleum reserves except the 2% or so in Indonesia until the recent oil discoveries in China.

Production in China has increased rapidly since 1970, and by 1975 approached that of Indonesia. South America, with one-tenth of Asia's population, has about as much oil, most of it in Venezuela. Africa, with about the same population as North America, has twice our known oil reserves, and that continent has an almost negligible rate of consumption.

These several comparisons make the position of the Near East an even greater paradox. Standard Oil of California established the first Saudi Arabian oil well in 1938, and they were soon joined by Exxon, Texaco, and Mobil to form Aramco. Starting with a rugged frontier existence, they gradually built rigs, pipelines, harbors, and towns. Saudi Arabia collected $10 billion from this foreign development before 1972. With well over one-half the world's known petroleum reserves, and production that was developed almost entirely by high-technology foreign companies, they have a political and economic weapon, oil flow, that can bring almost any nation to its knees. It is also very vulnerable to terrorist sabotage. Can America afford to depend on this kind of supply? Until very recently, the most common prediction about meeting future U.S. market demands was that, by 1985, 50% of our oil would be imported. What would be the eventual result of, say, $30 billion per year in oil imports that must be paid for either by exports of goods or by transfer of ownership of assets in like amount to others? There are plenty of reasons for us to develop more of our own oil resources and to use all our oil economically and wisely, with minimum waste. This is an urgent matter for early and continuing attention.

2. Natural Gas

This is a special case because the laws of supply, demand, and price have not been allowed to operate for this fuel in the United States. The price of natural gas has been kept at a low level for many years by federal regulation. This has gradually resulted in two effects: (1) More and more users shifted to natural gas, especially after antipollution regulations were enacted. Natural gas requires no antipollution devices; it is the cleanest fuel, uses simple combustion equipment, and no storage is necessary. In many locations it became cheaper than either oil or coal. Thus it was the preferred fuel for new or enlarged plants, for all uses. (2) With a low fixed price for their product, the industry could not attract new capital for prospecting. Natural gas suppliers had little incentive to prospect for and develop new sources because their available capital was needed for new pipelines, compressor and distribution stations, and service mains, at increasing costs of construction.

Even before 1972 these policies had led to a tight supply situation and looming shortages of natural gas, in spite of importation from Canada. Companies began to make large commitments for importation, including the development of liquid natural gas (LNG) plants, cryogenic tankers, and the revamping of facilities to convert this supply to pipeline gas. The future of the U.S. natural gas supply may be predictable, but the timetable remains very uncertain. If and when consumer prices are allowed to respond to demand and supply, LNG will become more attractive and additional domestic sources will also be found. The two uncertainties are the problem of paying for large imports and the fact that known U.S. reserves are very limited in comparison with the potential demand.

Recent estimates of U.S. and other reserves of natural gas are included in Table 13.[22] This table is a composite of estimates made by various authorities for known reserves of the three current fuels, natural gas, oil, and coal. Simple arithmetical projections are shown for the duration of these supplies (static life) for the U.S. and for the world. In other words, the apparent static life is the future duration of the supply if there is no change in estimated reserves, techniques of extraction, or rates of consumption.

A technical examination of estimates of reserves and of the effects of methods of extraction is highly complicated, and hardly necessary here. Official estimates of U.S. oil and gas reserves were revised downward in 1975. Comparisons are difficult to understand at a glance, since the units used vary from one table to the next. The first three items in Table 13 indicate the estimated U.S. reserves, expressed in three different sets of units for convenience. Table 13B gives 1972 consumption and reserves, both in the same units, from which the apparent static life as of 1972 is obtained by simple division.

Whatever method is used for defining the resources, present indications are that severe shortages must be expected sometime during the 1980s. The time and duration of the shortages will depend on developments in the meantime involving fuel conservation and use, as well as exploration and extraction. It should be noted from the last line in Table 13 that even the known supplies of petroleum and natural gas in the entire world are very limited, and international attention to conserving them and to the development of substitute energy sources is important to all peoples.

Natural gas is the most convenient fuel for nontransport uses, and its supply deserves attention equal to that being given to the oil shortage.

---

[22] Table 13. World Fuel Resources, 1972: Various Guesses on Recoverable Fuel Underground.

It is apparent that most estimates predict exhaustion of the known U.S. reserves of natural gas by about 1990. Doubtless the exhaustion will be delayed as natural gas or its equivalent becomes available from three developments already in prospect: (1) gas from coal, preferably with enrichment to approach the heating value of methane-type natural gas; (2) gas from other hydrocarbons and from organic wastes; and (3) hydrogen from electrolysis of water, probably by using nuclear energy in remote plants. All these methods are already in experimental use, and they are considered separately in later sections. It is evident that supplies of natural gas are even less plentiful than those of petroleum liquid fuels. Substitute fuels will be expensive, and large substitute supplies will probably not be available in the next 10 or 15 years.

Recent demands for natural gas are perhaps 20% in excess of the supply. If strict well-head price controls are continued, the shortage will doubtless increase. If prices are gradually decontrolled, the demand will level off as prices climb; but high oil prices make gas more attractive, and at even twice the 1970 price the demand for natural gas will probably continue to grow.

It has been suggested that large new supplies of natural gas may suddenly be discovered when extensive prospecting is undertaken. Nevertheless, conservation measures are an urgent precaution. They are needed soon. A most radical proposal is that, since natural gas is such a valuable raw material for the chemical industries, the entire present supply ought to be reserved for this purpose, and its use for fuel practically eliminated. Any such changeover would probably not be tolerated by today's gas users.

All present indications are that some of the alternative measures ought to be implemented at an early date, at least as a part of U.S. energy planning for the 1980s. For the balance of the 1970s, the prospects are for higher prices of natural gas and a small increase in the supply or in the proven reserves. The percentage of total fuel energy supplied from natural gas probably will not (and should not) ever again be as high in the United States as in the recent past, but its advantages as a fuel will continue to bolster the demand, and the supply.

There is vigorous argument about the future of natural gas in the United States, and the outcome will not be known for several years. In view of the present wide distribution of gas, and the investment in pipelines, compressor stations, distribution and metering systems, appliances, furnaces, and gas-heated equipment, it is virtually unthinkable that high-Btu gas will disappear from the market after the approximately 15 years underground reserves are expected to last. In the first place there is much disagreement about the extent of the reserves. The

270 trillion cubic feet of Table 13 is close to the American Gas Association (AGA) value, which has been reported yearly for some time. The FPC recently used the figure 1200 trillion cubic feet for the ultimate reserves, and the Federal Trade Commission (FTC) has been trying to ascertain the accuracy of the industry data. The public resists rate increases, and this will doubtless increase scarcities temporarily, after which prices must rise to a more reasonable level. Gas from coal and other hydrocarbons, LNG (imported), and hydrogen will then become competitive, and commercially available. The entire development is an interesting struggle but, like the automobile, the gas furnace and the gas appliance are here to stay.

3. King Coal

Coal is the great American energy resource. It could support our entire energy economy for hundreds of years as far as gross Btu supplies are concerned. Up to about 1870 coal furnished less than 20% of all the energy used, with firewood supplying most of the rest. By 1910 the fraction furnished by coal had reached almost 80%; but as cleaner and more convenient fuels became available, the demand for coal declined almost as fast as it had risen; and by 1970 the 20% figure was again reached, with the relative demand still dropping. Now is the time for a fresh look at this great resource.

Total U.S. underground coal deposits are estimated at a trillion tons or more. With production at about one-half billion tons per year, this means an ultimate supply for about 2000 years at present rates of use. Yearly production must be much more than doubled, however, and very soon, if coal is to provide major savings in petroleum and natural gas.

There is much disagreement about the proven reserves of coal in the United States. Several recognized authorities have made recent estimates of our ultimate resources, setting a total at least five times as great as the 500 billion tons of proven reserves given in Table 13. About one-half of Eastern deep-mine coal is left in the mines, for reasons of economics and safety. Most Western coal and lignite is low grade, high in water and ash content. Some newer fields have not been well explored, and their extent is somewhat of a guess. Mining practices are changing rapidly, increasing the percentage of the total coal that can be extracted. Other great advances are in prospect for the next few years.

The production of gas and oil from coal was a live subject 50 years ago; and many ideas were tried in experimental plants, some of which were of considerable size. It has been estimated that significant oil and gas production from coal will be a reality by 1990; and eventually this might involve the added use of billions of tons of coal per year, perhaps tripling the demands on coal mines. Since such operations are still in the pilot-plant stage, an accelerated time schedule will be necessary.

Many factors will determine the real desirability and the time scale of such a program. It would probably call for mining more coal in the United States in the next 25 years than has been mined in all our past history (about 50 billion tons). In addition to the vast capital and manpower demands for such a project, further improvements in technique must be developed for mining at greater depths, with thinner fuel seams, with a higher percentage of total recovery, and using larger machines. Surface mining (strip and auger) has already exceeded underground coal mining, and it will contribute the greater share of the projected increase, mostly in Montana, Wyoming, and North Dakota, where about one-third of the most accessible coal deposits are found. These deposits are low-grade bituminous coal and lignite, however, and with overburden up to 150 feet or more, the costs of even modest control of pollution will be tremendous. The solid wastes from mining, the residues from processing these low-heat-value fuels for gas and oil production, the volatile products wasted to the atmosphere in any conversion process, and the water required pose technical problems in each step of the process, including pollution control. Materials must be handled on an unprecedented scale. The final result will be a costly product which could only compete with high-priced gas and oil. The economic balance will be worked out gradually, but the former 1980 target date for significant large-scale conversion of coal to gas and oil is entirely too optimistic.

In the meantime, the much wider use of coal in present and future central-station electric power plants is one of the most promising ways to save gas and oil. Reconversion of plants formerly equipped to burn coal, but shifted to gas or oil after the Clean Air Act, is a first step. But here, as with new plants, the high cost of adequate pollution-control equipment raises some questions. If desulfurization of coal became economically feasible, major equipment for the removal of sulfur oxides from stack gases will not be necessary. These contrasting techniques involve economic problems, as well as new technology. Long-term planning is necessary because lead times are great, so the difficulties are evident, and the final result will be higher-cost power, with resultant effects on all consumer uses.

Large-scale increases in mass transportation to save gasoline are likely to reinstate some electric rail lines and trackless trolleys, increasing utility loads. Electric automobiles would use utility power for battery charging. Summer air-conditioning loads are already the peak power users, and are still increasing. This makes the use of high-level electric energy for low-level heating, for residences and small buildings, attractive to the utilities, because installed generating capacity is available and thus the high fixed costs can be spread more widely.

All these uses transfer some of the energy load to coal. Large transmission losses are involved. The use of high-level electric energy (i.e., high available energy) for low-level heating and heat pumping (mechanical cooling) is technically abhorrent, and a large argument persists as to whether electric heating for comfort should be prohibited. However, it saves gas and oil, lowers building and equipment costs for residences and small buildings (no chimneys, for instance), and saves cleaning and maintenance costs.

In spite of great coal resources and the feasibility of large-scale mining there is still a question whether the highly desirable results will become unattainable within the critical period of the next 25 years. There is a need for public education, so that decisions can at least be made with full knowledge of the facts about the need and the environmental threats. Possibly, the great strides that have already been made in the control of pollution in our country may be endangered or partially abandoned through the rush to better utilize our great reserves of energy in coal.

Summarizing the American scene as regards coal fuel and its future prospects: We have plenty of coal; we have been using it only slowly; and now there is every reason to accelerate its use. Electrical utility power, based largely on coal, has been growing more rapidly than other forms of energy, and it is agreed that this must continue (see Section B and Tables 10 and 11). There are two inhibiting forces, namely, the very high capital demands and the environmental difficulties. Doubling the electrical output well before the turn of the century seems most desirable and even necessary, but this means at least doubling the present total investment in utility plants. Coal production and the attendant investment must grow even more rapidly, because more coal will be used not only in utility plants but as a source of synthetic pipeline gas and oil. More coal will also be needed in the steel industry when the massive construction program covering the entire energy field becomes a reality. Counter to these expanding forces are large environmental problems, including mine safety and large manpower needs. As to the latter, while it is true that both coal mining and electricity production are becoming less and less labor-intensive for their operation, expansion does call for large technical forces for design and construction. Technical manpower will be in very short supply (Chapter 7, Section B).

The public, through its effective task force of environmentalists, will demand the solution of pollution problems arising out of the expansion of power plants and coal mining and the building of an industry that produces synthetic gas and oil. Among the difficulties are atmospheric pollution, especially by sulfur oxides, land restoration after strip mining, mine safety underground, water shortages, and thermal pollution of scarce water supplies. New technology will be called for, but the most glaring difficulty will be *delay*. Before construction, endless environmental impact reports and hearings are demanded. If, or where, environmental criteria are not met, it will be "back to the drawing boards." This necessary but slow process will make it very difficult to meet the growing electric and fuel energy needs.

Thus the great possibilities for using more coal instead of gas and oil in the continuing energy crisis may be fulfilled rather slowly, even though all steps are feasible from the standpoints of resources, technology, and economics.

In the longer and wider view involving the use of coal as a greater energy source worldwide over the next 100 years, one result must be the great expansion of underground mining, demanded by the locations of known coal reserves. In the past, the safety records of underground mines have been poor. While there has been much mechanization and safety improvement in recent years, in most countries the reputation of the mining industry as regards progressiveness leaves much to be desired.

Although the United States has a fair share of the world's known coal reserves (see Table 13), most countries in the northern hemisphere, including the Communist countries, have sizable coal reserves and a developed coal industry. Thus, in the industrialized countries, coal is in a good position to supplement oil and gas; and there is every reason for international cooperative efforts in that direction. The extent of that cooperation will depend on the attitudes of coming generations, and on the future expansion of nuclear energy for both constructive and destructive purposes. A crisis may also arise when the developing nations of southern Asia, Africa, and Latin America greatly increase their demands for a larger share of the world's energy supplies. Scarcity of coal reserves in the southern hemisphere will then become a significant factor.

4. *Hydroelectric Power*

Hydropower is basically solar energy. In the United States it furnishes something less than 4% of the total energy supply (Table 12). The best locations have already been developed; but with sufficient investment, the total capacity could probably be doubled. There is little prospect of increasing the total percentage contribution, but a well-planned program for the critical 25 years just ahead would at least enable hydroelectric power to be increased in proportion to the total energy consumption. As fuel prices increase, several of the present undeveloped water power sites will become economically more attractive, because hydropower uses no fuel and produces no pollution.

Any large hydroelectric project involves many purposes in addition to power generation. Downstream areas can be largely protected from floods by lake storage above the dams, and lakes offer wide recreational advantages. Groundwater reserves may be increased. Pumped storage during off-peak periods allows power generation between an upper and a lower lake during peak loads. Operating costs and maintenance for the whole system are low in comparison with fuel-fired or nuclear plants.

However, a large lake inundates towns, roads, and farmlands, thus destroying certain community values and taking scarce agricultural land out of use. For maximum power and efficiency the water level in the upper lake should be high, but the lake then could not contain a heavy flood. During an extended drought the lake level would become too low. Power would then be at a minimum, and recreation possibilities unattractive or even nonexistent.

Many prospective hydroelectric sites are in sparsely settled territory, and electric power transmission to industrialized area would be too expensive with present technology. There are, however, several rivers and tributaries where local demand for power is good and flood-control measures are needed, for example, the Susquehanna, the Ohio, and the Mississippi. In view of the energy shortage such projects can be justified on the grounds of public policy in spite of the large investments required. Three-fourths of the undeveloped capacity is in the Far West, mostly at some distance from population centers.[23]

Hydroelectric power outside the United States was about twice that generated in this country in 1973. Almost thirty large hydroelectric plants, 1000 megawatts or larger, were under construction in 1973, with an ultimate capacity of over 60 million kilowatts. One-third of this number were in the USSR (capacity 25 million kilowatts and only 4.3 million kilowatts in the United States). The total potential water power in Africa is reported to be four times that of the United States, and little native fuel is available except North African oil. Worldwide construction of dams to provide hydropower appeals to the public imagination. In 1973 almost fifty dams were under construction or just completed. The typical prob-

---

[23] Table 14. Electrical Utility Power, 1950–1974: Capacities and Outputs from U.S. Fuel, Hydroelectric, and Nuclear Plants.

lems of transmission and utilization of power are obviously present, so the overall contribution to solving the near-future energy shortage will be small. These resources will no doubt be of more concern at a later date. Nevertheless, every economical contribution to power supplies today should be welcomed and expedited.

5. Nuclear Power

"By far the greatest promise for increasing the available supply of high-grade energy is to be found in the field of atomic or nuclear power." Such a statement has been repeated time and again since the first nuclear power stations began operating in the 1950s, and there continues to be overwhelming evidence to support it. Nevertheless, nuclear capacity has been developing slowly. It took almost 20 years for nuclear-source electric power to reach an annual output of 1% of the total generated. Worldwide, in 1973 before the economic recession upset the growth pattern, nuclear power represented about 3% of total electric power generation. Almost two-thirds was produced with light-water reactors, and one-half of this by pressurized water reactors (PWRs). Gas-cooled reactors provided a little less than one-fourth of the total, and heavy-water reactors less than one-tenth in 1973.

The British took an early lead in nuclear power, using gas-cooled reactors. The United States has installed water-cooled reactors, and in the 5 years preceding the economic recession a large number was installed or exported. By 1974 there were over fifty nuclear plants in commercial operation in the United States, and another 200 on the drawing boards or under construction. Almost one-half of the latter were subsequently postponed or canceled, largely for economic reasons. But uncertainties about the future of nuclear power began to increase. Major campaigns were launched that questioned the safety of nuclear plants, fuel, and residues. Relative costs rose sharply, and plant siting was delayed by local groups.

Proponents of nuclear power failed to pay enough attention to this rising tide of opposition based on the safety issue. Objectors became more vocal, both in the United States and in certain European countries, notably Switzerland and Sweden. With both increasing costs of plants and serious delays in attaining full-load operation, even investors began to question the future of nuclear power. However, both President Nixon and President Ford had advocated large federal research and development programs heavily weighted toward nuclear power.

Small but highly organized groups such as the Union of Concerned Scientists and the Consolidated National Intervenors conducted effective opposition campaigns. Congressional debate loomed after President Ford vetoed the bill extending the program of federal insurance against nuclear accidents. Ralph Nader called for an outright moratorium on nuclear power development and was quoted as saying: "If people knew what the facts were and if they had to choose between nuclear reactors and candles, they would choose candles." Early and inadequate simulation tests were quoted by the detractors, and the supporters of nuclear power did little to furnish better information. For example, back in 1957 a study at the Brookhaven Laboratory on a hypothetical reactor of 100- to 200-megawatt size, presumed to be located near a large city, resulted in the prediction that complete failure of the reactor and *all* safety devices would kill 3400 people, injure over 40,000, and contaminate an area the size of California. Other tests in 1971 in Idaho are also cited. In these tests a small model of the emergency core-cooling system (ECCS) was used with a simulated reactor which consisted of a 9-inch-diameter drum electrically heated. Failure occurred in all six tests conducted.

The ECCS is a major safety device in all water-reactor nuclear power plants. At four different plants, during 1974–1975, minute cracks were discovered in the ECCS piping systems. This aroused suspicion about the reliability of these vital emergency systems, and twenty-four reactors were ordered shut down for careful inspection and tests. In October 1974, at one of the nuclear plants of the Commonwealth Edison Company a pump was shut down through human error and a sensing device energized the ECCS, but it failed to function. The reason reported for this failure was that three different valves did not function, two failed to open and one failed to close. There have been many small failure incidents in the 2500 or more reactor-years of operating experience to late 1975, including failures of controls, flow blockages, leaks, and bearings scored, but practically all these failures were in nonnuclear areas, and no failure ever resulted in even potential injury to the public. Such troubles occur in any power plant, and Commonwealth Edison has in no sense abandoned nuclear power. They report that one-third of their generating capacity is nuclear and that the cost of generating power in these plants is less than with coal. In November 1975 one-half of their electric power output was from nuclear plants.

Early objectors to nuclear power aroused the public by such statements as, "A reactor contains enough fissionable material to make hundreds of atom bombs. Do you want *that* in your backyard?" More recently another group declared that "lethal reactor wastes call for fail-safe storage for five hundred thousand years, during which there must be no social instability or earthquakes." The facts are that there have been no accidental deaths or major injuries at any nuclear plant. Safety measures have exceeded almost anything in the past history of any and all energy industries, and amazing safety records have resulted.

Two major U.S. studies on nuclear power safety were the MIT report in 1974 and the Nuclear Regulatory Commission (NRC) report in 1975. The earlier report, comprising fourteen volumes, was issued by a team directed by Professor N. C. Rasmussen of MIT and is entitled *An Assessment of Accident Risks in U.S. Commercial Nuclear Power Plants*, (WASH 1400 NUREG 75/014). It noted that there have been *zero* fatal accidents in 100 nuclear plants. It estimated that an individual's chance of fatality in a motor vehicle accident is over 75,000 times as great as that from a nuclear accident, and that the risk of injury from some other kind of accident is 10,000 times as great as the nuclear risk. The NRC report estimated that the chance of a person living near a nuclear plant being killed by a nuclear accident was one in five billion per year. Large numbers are not too meaningful to the public, but Chairman William Anders of the NRC stated that plants that conform to NRC's requirements provide adequate protection to public health, safety, and the environment. The contents of these reports should be studied, because their extensive evidence is reassuring.

The advocates of more stringent measures for nuclear safety have been joined by several scientists and engineers who had previously held important positions in the nuclear power field. Their presence among the opponents of nuclear power was widely publicized and aroused public suspicion that nuclear safety was still a serious and unsolved problem.

Adding to the rekindled concern on the part of the public for the safety of nuclear power was the serious fire at TVA's Browns Ferry (Alabama) nuclear plant in March 1975. The fire was started by a carelessly used inspection candle which ignited insulation on certain safety control cables. It finally resulted in extensive damage to two reactors. The official 1976 report on the fire concluded that programs of fire prevention and control for this installation had been "essentially zero." The report called for a general improvement in NRC programs for the prevention, detection, and control of fires in both new and existing plants.

Fire danger exists in plants of every description. In this plant it had apparently not been given adequate attention in either the original design or in the "quality assurance" inspection programs during plant construction.

Public concern over nuclear safety and over environmental protection in general was much in evidence during 1975–1976. Several examples such as the Browns Ferry fire, the Alaska pipeline difficulties including the questionable pipe welds, and the failure of the Teton Dam, led to a wide conviction that major industrial and construction projects are often undertaken without adequate consideration of the interests of the public. The zealous opponents of nuclear power quickly turned this concern to political advantage.

A direct-initiative proposal (No. 15) was placed on the 1976 California ballot that favored drastic nuclear safety measures. It was widely held that this proposed law would eventually prohibit nuclear plant operation in California. After an intensive and massive campaign, with scare tactics used by both sides, the proposal failed to pass. But this was only a first attempt, and since at least 20 other states have direct-initiative laws, and environmental lawmaking by this means was already successful, some nuclear plant regulation by popular initiative may be expected. Highly technical questions are involved in the nuclear safety issue however, and legislation by popular acclaim and emotional argument is a poor substitute for technical evaluation and serious legislative debate and compromise.

Proponents of nuclear power point out that absolute safety is unattainable in *any* energy system, and that such a demand would have completely prevented the development and use of automobiles, airplanes, television sets, and even home-heating furnaces. The ecologist counters that we should no longer allow the large-scale development of *any* of the products of science and invention until all possibilities of catastrophic damage to life and environment have been thoroughly investigated. Such a demand probably would have stopped the installation of the first large machine shop. But many people regard this as a tenable position; and the resulting argument is likely to continue to delay the rapid expansion of nuclear power in the United States, if for no other reason than the wide suspicion that government nuclear agencies are not telling all they know about radiation dangers and nuclear safety. Zero radiation dosage is admittedly unattainable; but living at a high altitude, in Denver, for example, statistically provides greater radiation exposure than any planned expansion of nuclear power.

For several good reasons, a light-water reactor cannot act like an atomic bomb, no matter what the conditions. Reactor fuel is only slightly enriched in fissionable material, as compared with the natural product, and its final percentage does not approach that of a bomb. Reactor fuel has constituents that would prevent a bomblike reaction anyway. Its surroundings also contain neutrons, while a bomb must be set off in "clean" surroundings. A sudden firing sequence, in millionths of a second, is used to set off a bomb, and nothing comparable is possible in a reactor. The bomb argument has been pretty well laid to rest, but what about the escape of radioactive material, perhaps following some failure of the cooling system, or even during loading or transportation of fuel?

No manufactured device or human operation is perfect, but a reactor tends to shut itself down if overheated and, while other failures are at least possible, they are not at all likely to result in spillage of radioactive material into any part of the system. Even if such spillage should occur, there are flushing and chemical means of cleanup and transfer of the dangerous material to existing holding tanks. There has never been a reactor accident that ruptured the main reactor drum or tank, which is of special steel and about 10 inches thick. In any event, the very dispersion of the molten material into a much larger area and volume would lower its extreme temperature and soon arrest its spread so that it did not impinge on the main tank.

Outside this main tank is an airtight dome of steel and reinforced concrete, probably 4 feet thick, which is all the public sees. It encloses all the nuclear components including the steam generator, piping, and pumps. This shell is designed to resist damage from any imaginable accident, but there is of course no way to test it full scale. Inside, pumps and other safety equipment are installed in duplicate, usually with three alternate sources of power, under automatic sequence control.

Each of the several barriers to escape of radioactive material is of such strength, size, and mass that, according to calculations, it cannot fail, and the statistical probabilities of failure are miniscule. No conditions are perfect, but the safety precautions in a nuclear power plant are unique and known in no other industry. A favorite evaluation is that "the hazard is about the same as the hazard of being struck by a meteor." Safe procedures for transporting nuclear fuel and wastes have been developed, but human error or a major accident is always possible. The same can be said for the operations of fuel preparation and processing. No industrial operation is entirely safe from error and accident.

Spent reactor fuel and waste are radioactive, and breeder reactors contain plutonium, a raw material used in atomic bombs. What is their possible danger to present and future generations? One safety advantage, not usually appreciated, is the small volume of the reactor fuel and waste. A fuel quantity the size of an aspirin tablet would provide electricity for an average home for a year (7000 kilowatthours). Coal to generate this much power would fill sixty bushel baskets. One authority has stated that if the entire present U.S. electric power generation were converted to nuclear power and then run for 350 years at this level, the nuclear fuel waste would amount to a 200-foot cube, which could be stored in a single small mine-type underground space. In comparison, if such an operation were entirely fueled by coal and the ash and refuse dumped into mines of similar size, about 75,000 mines would be required. There is no question but that reactor wastes will continue to be dangerous for many human generations. Some authorities suggest 350 years for safe storage time, but as long as fission reactors are used there will continue to be new wastes, of maximum radioactive strength, in storage. Of the radiation waste in storage in 1976, however, less than one-tenth is from power plants, most of the rest having resulted from nuclear weapons programs.

An argument against extensive expansion of the use of present-day reactor designs is the high capital outlay that would be necessary if additional plants for uranium enrichment are necessary (i.e., from about 1% uranium-235 to 3% uranium-235). If the great diffusion plants built by the United States for atomic bombs, costing billions, and the European centrifugal separation plants, became inadequate, costly additional plants will be necessary. These will not be needed after the breeder reactor becames the accepted commercial design.

Since electric power will continue to be generated by one means another, the question of *relative* danger is the significant one. A study reported in *Nuclear Safety* in 1964 (C. Starr, "Radiation in Perspective") showed that 19,000 deaths a year in the U.S. were directly attributable to the mining, processing, and use of coal and oil. At this rate, if 10% of these fuels was displaced by nuclear energy in the near future, 1900 deaths per year would not occur, or 47,500 deaths in an assumed 25-year life of a reactor. Since there have been few deaths chargeable to uranium mining and none to reactor operation in the 15 years of reactor power, the comparative safety record of nuclear power is certainly very good, even though the process is new.

Even if the safety argument can somehow be resolved, there are narrow limits to the rapid expansion of nuclear power. For one thing, every nuclear power plant to date has been unique in design and construction, resulting in a lead time of 10 years or more from site selection to full commercial availability. Few plants have been able to start construction in less than 4 years, and many completed plants have not been allowed to operate at full-rated load. There are now strong attempts to improve this by standardization of equipment. Site locations could be classified and permits granted concurrently for several plants to be located on sites of a specific class. Before long it might then be possible to reduce the lead time to 5 or 6 years, comparable to that for a large fuel-fired plant.

The possible illegitimate use of nuclear materials for terrorism or violence is one more danger in the pattern of lawlessness to which modern society is increasingly being subjected. Since a 5-pound charge of fuel in a single fuel rod has the energy potential of 6000 *tons* of coal, the material is more valuable than gold or jewels, and worth stealing. Unprincipled possessors of nuclear material, whether individuals, gangs, or ambitious tyrants, are indeed a danger to humanity, but stealing and making use of the material is not as simple as it sounds. Radioactivity is extremely dangerous, and a thief must know how to protect himself. Unlike gold or jewels, there is a very small market for nuclear fuels. Plutonium is lethal, and using it in making a bomb calls for highly sophisticated techniques quite unknown in a semiprimitive society, plus facilities that exist in only a few locations in the world. Nevertheless, nuclear materials *are* possible tools for terrorism, blackmail, and war, and the more locations in which they can be found the greater the dangers of their illegitimate use. If a society cannot control the lawless use of guns and explosives, the same will be true for nuclear materials, with the limitations only those of accessibility and knowledge of use. The ultimate dangers to the human race, however, however, are very small as compared with the overkill stocks of nuclear devices which are kept growing through the use of our yearly taxes.

In spite of all the difficulties, there have been many predictions that our energy needs will outweigh them, and well they may. A favorite assertion is that nuclear electric stations will carry over one-half the total load in the United States before the year 2000. Even if this were possible, it would call for answers to several questions: Do we have enough uranium? Where is there enough cooling water for so many plants in view of the great amount of heat wasted by the present inefficient reactors? Would it not be much better to concentrate on the production and use of coal for power until the breeder reactor has been developed to commercial status, or even to concentrate on earlier success for the fusion reactor? Like most energy questions these must be considered with due regard to the related environmental factors. They also involve tradeoffs as regards capital investment and manpower utilization. Extensive computer studies are indicated, since so many variables are involved.

Whatever the pattern of increase in reactor-generated power, the human and ecological dangers will persist until the fusion reactor becomes a reality, after which new ones will doubtless take their place. The fusion reactor appears almost as distant today as it did a generation ago, and most predictions are that the present living generations will not see this accomplishment fulfilled. Thus, even though there are many intriguing possibilities, fusion reactor development is outside the scope of this discussion.

D. ENERGY SUPPLIES FROM FUTURE SOURCES

Synthetic oil and gas from coal, shale, and tar sands cannot be produced in the near future at prices competitive with petroleum and natural gas. The same is true for substitute fuels such as methanol and hydrogen, as well as for solar power. High capital costs are involved in each case, and equipment for large-scale production has not been developed.

If energy independence is to be attained in the 1980s, some forms of selective federal subsidy will probably be necessary. If increased taxes were placed on all common fuels for financing such subsidies, the resulting higher prices of the common fuels would make the expensive new fuels more competitive. However, this would add to the problems of inflation, customer resistance, and social unrest.

1. Synthetic Fuels from Coal and Oil

The production of gas fuel from coal and oil is an old story. In fact, it is reported that, as early as 1825, London had 125 miles of street mains for the distribution of coal gas. By 1925 Britain had 40,000 miles of such mains, serving seven million customers. By 1900 most large U.S. cities were supplied with coal gas from local plants, largely for lighting and cooking. As late as 1920 little U.S. natural gas was available through pipeline distribution, but after World War II the pipeline network grew very rapidly and coal gas was no longer used.

Most of the manufactured coal gases, including coke-oven gas, contain toxic carbon monoxide and have a disagreeable odor. Pipeline coal gases have a heating value of about 500 Btu per cubic foot, producer gas less than 200, and blast furnace gas less than 100. Both retort gas and coke-oven gas are made by high-temperature processes (above 1500°F), but there has been much experimentation with "low-temperature carbonization" (say, 1100°F), which gives coke with about 5% volatile matter and gas of calorific value as high as 900 Btu per cubic foot.

Oil gas, from low grades of oil, has been produced by various processes. It has a high heating value (1000 to 3000 Btu per cubic foot) and can be used for enriching coal gas to attain the 1000-Btu value of natural gas. As the petroleum and natural gas industries developed, they also produced propane and propane-butane mixtures, with over 3000 Btu per cubic foot. These are sold as bottled gas for small isolated or portable gas-burning devices, including engines.

In spite of this long and varied background in the production of gas from coal and oil, recent markets have been very limited, largely as a result of the lower cost and wide distribution of natural gas and refined petroleum liquid fuels. Now, for Project Independence in particular, it is essential that large-volume production of both 1000-Btu gas and synthetic oil from coal be accelerated.

Improved processes for these purposes are being pushed by the Office of Coal Research, the Bureau of Mines, the AGA, several major oil and coal companies, and several other large corporations. Three products are in view for quantity production: pipeline-quality gas of about 1000 Btu per cubic foot, lower quality but cheaper power-plant gas, and synthetic oils for supplementing the available supplies of petroleum fuels. At least five processes are now in the advanced pilot-plant stage. Most of the newer developments use multistage gasifiers, and they turn out more than one product. Quantity production of cheaper, low-Btu coal gas, for a high-efficiency cycle utilizing both gas turbines and steam turbines, is one promising arrangement. It can be expected to raise the efficiency of the best electric power plants to almost 50%.

Much capital will be needed, as this production of synthetic fuel will be a whole new industry (see Chapter 7, Section A). Gigantic materials-handling operations will also be involved, first in handling the strip-mined coal and its overburden, then in crushing, pulverizing, retorting, or processing the coal in fluidized-bed reactors, and finally in cooling, condensing, scrubbing, and separating processes, and disposal of residues.

As of 1975 the major undertaking of large-scale synthetic fuel production from coal was just starting, amidst much questioning and many objections. The development of "cowboy coal" industries in the West meets with objections from environmentalists and even from state officials there. The Office of Coal Research of the DOI awarded a contract of unprecedented size ($237 million) for a demonstration plant, with the expectation of necessary industry participation. It was reported that the plant might be in operation before 1980. This contract follows a half billion dollars worth of government support for coal conversion projects over many past years. Smaller plants have been built and operated, both in the United States and in Germany and South Africa; and this provides very necessary experience for this large demonstration project. American industry is actually highly involved already, in both the mining and the coal conversion projects. The Burlington Northern Railroad owns land under which it is claimed that a coal reserve of 11 billion tons exists. Five other companies own another 11 billion.

With five western states accounting for well over one-half of all U.S. coal reserves (North Dakota, Montana, Wyoming, Utah, and Colorado) the activity in the area is fast increasing. The unsuccessful experience with land restoration after strip mining in the eastern states is being cited, since it is presumed that all the western coal will be strip-mined. Concerned citizens and public officials are already objecting, and this makes the industrial participants apprehensive.

The scale of this operation would eventually call for about twenty-five or more plants, and a very high capital investment. This should provide the equivalent of about 10% of the natural gas requirements, plus an uncertain amount of oil, possibly by 1985. Such a program could prove to be an impossible order, however, since full-sized plants have not even been developed yet. Perhaps the capital and manpower might better be used in the immediate future for projects for which assured returns can be obtained more quickly, for example, more prospecting for petroleum and natural gas, and more electric power plants of conventional, coal-fired design.

In view of the uncertain schedule of mass-producing gas and oil from coal, an interim solution for the natural gas shortage would be to expand greatly the production of synthetic pipeline gas from petroleum liquids. Such an expansion would be limited both by the high cost of the product and by the urgency for reserving available petroleum for transportation and for the petrochemical industries. In general, available capital, manpower, and materials should not be used for interim or temporary expedients, but rather employed for programs of long-term significance only.

2. *Shale Oil and Tar Sands*

The tremendous worldwide deposits of oil shale and tar sands have enthused many a promoter. A fair amount of work on oil extraction from these materials has been done, not only in the U.S. and Canada, but also in the USSR, the Baltic States, Italy, and so on. There are major oil shale deposits in Asia and Africa. Most highly endowed is North America, with shale deposits in Colorado, Utah, and Wyoming, and rich tar sands in Alberta.

In comparison with OPEC crude oil at $10 a barrel or more, the costs of oil from either U.S. shale or Canadian Athabasca tar sand would be within the range of economic competition; but there are difficulties, the major one being that processes and plants on any major scale have not yet been developed. Because of the lead time required for such development, all the major studies dealing with the energy crisis and what to do about it have concluded that these sources of oil will be insignificant for several years.

The NSF-RANN report of 1971 (footnote 16) gave the estimate that world reserves of shale oil are equivalent to nearly 20,000 times the current annual world consumption of energy, and stated that more than two-thirds of the proven reserves are in the United States. Several pilot-scale plants have been set up and operated, and both the U.S. government and private corporations have financed extensive studies on the design and operation of large-scale facilities.

One of the unsolved problems is that the shale oil recovery process uses a large quantity of water and most of the water sources in the shale region, for example, the Colorado River, are already in full use. Air pollution and solid-waste disposal are other deterrents to large-scale operation of shale oil plants. Pressure roasting of whole mountains of rock is indeed a vast project of materials handling.

Shale oil has been assigned a rather low priority in research and development planning. The comprehensive program projected by the Energy Policy Office and reported to the president by the AEC in 1973 (footnote 17) covered all forms of energy research and development, with total federal participation of $10 billion in a 5-year program. Shale oil projects were alloted only 1.3% of the total, as compared with ten times as much for programs for developing synthetic oil and gas production from coal. In comparison, breeder reactor development was set at an expenditure level more than 100 times that for shale oil. ERDA has since changed the priorities somewhat.

There is every indication (in 1976) that research priorities will continue to change rapidly in response to developments during the next 5 years. These controlling developments will include not only the relative successes in research and pilot-plant operations, but also the success of fuel conservation programs and decisions on environmental protection. Although there is great objection to nuclear power development of all kinds, based on the seriousness of possible failure or sabotage, no substitute for the large-scale power output of nuclear plants will soon be available. In any consideration of shale, solar, or any other new source for the large-plant power output we need, the certainty of massive pollution from such new sources must be weighed against the extremely remote possibility of a nuclear accident. The great quantity of unavoidable pollution from shale or coal and landscape pollution from solar or wind power, become the alternatives to the billion-to-one chance of a nuclear accident or future problems with nuclear waste.

It is not to be expected that the United States will abandon its high-energy living standard. It seems very likely, however, that the delays in deciding which of our possible sources is to become the main substitute for oil and gas will continue to the point where major black-outs, shutdowns, and traffic stalls will be common. Be it shale, solar, nuclear, or coal that eventually becomes our mainstay, the time required for full development is measured in decades, and we have already lost several years in getting started.

Tar sand is a minor energy source which invites parallel develop-ment with shale oil because of certain similarities. Canadian deposits are the major sources at present.

Athabasca tar sand is a water-oil-sand mixture with about 10% oil by weight (more by volume). These sands are mostly about 50 feet below the surface, easily accessible by strip mining most of the year. But the frequent temperature of 30° below zero in northern Alberta in winter does cause some difficulties, because of the water content and the great quantities of water and steam involved in the usual oil reclaiming pro-cess.

The Great Canadian Oil Sands Project (Sun Oil) has been operating since the early 1970s. At first the Alberta provincial government limited it to a production of 45,000 barrels a day, later raising that to 65,000. Four more plants either have been started or are planned, involving a half dozen companies in all. The provincial government tries to control the economic struggle between tar-sand oil and the great petroleum output nearby. Over 80% of Canada's petroleum production comes from Al-berta. The government freeze on the price of oil from tar sands has caused a profit squeeze, resulting in large net losses.

The sulfur content of tar-sand bitumen is high, but the GCOS process recovers it as a by-product and produces an almost sulfur-free crude which is in high demand. It is reported that production and capital costs of the final oil product can be held to the range of $3 per barrel (1970 dollars), but the difficulties of a new process combined with a govern-ment freeze on the price of the product allowed no room for profit in the first several years.

One authentic estimate of the Athabasca tar-sand reserves indicates that, if the four plants now projected eventually attained an output of half a million barrels per day, they could operate over 1000 years at that rate and use only one-half of the Athabasca tar-sand reserves.

The United States also has small tar-sand deposits, mostly in Utah, estimated to contain about 20 million barrels of oil.

Oil from tar sands is a small but valuable addition to oil production in North America. Both shale oil and tar-sand developments are challeng-ing. They should and will be developed as rapidly as economic condi-tions permit, but it is likely that for some time their contribution will be less than 5% of the oil used in America.

3. Breeder Reactors

In the report, "The Nation's Energy Future," submitted to the president in December 1973, by Dr. D. L. Ray, Chairman of the AEC, a $22.5 billion energy research and development program was recommended for 1975–1980. Of this total about 55% was expected to come from private sources. In the detailed program which outlined each of five major tasks, the largest task called for $5.34 billion to "validate the nuclear option." Over 75% of this was expected to come from government funds. With the recommended research and development programs, nuclear energy was expected to furnish by 1985 about 15% of the total U.S. energy, compared with 27% from coal, 53% from gas and oil, and less than 4% from solar, geothermal, and hydroelectric sources. About 95% of the federal funds for nuclear energy were to be spent on the development of the breeder reactor. Prior to this 1973 report the breeder reactor had long had first place in energy planning.

The massive 1971 study, *The U.S. Energy Problem,* by NSF-RANN, anticipated that nuclear power would furnish one-third of all electric power in 1985, about 40% of this from breeder reactors, with all other reactors retired before 2030. The 1974 report to the Club of Rome carried a "nuclear option scenario" partially based on AEC data, showing 65% of all energy supplied from nuclear plants by the year 2025.

This high regard for the breeder reactor seems to be waning somewhat in the face of great opposition from environmentalists. The program is already years behind the earlier schedules, and there is rumor that the first plant will cost $10 billion and not be in operation until 1990. In the meantime the French have built a 250-megawatt plant, reported to have cost only $140 million. This plant has been operating since early 1974, with an availability of 84% (1976 report). There is a similar plant in Scotland. Russia started operating a 350-megawatt breeder in 1973 and is building a 600-megawatt plant. Several others are planned, with Italy and West Germany included. The ones in Sweden and Switzerland have been postponed by environmental questions.

A breeder reactor plant is more complex and expensive than a nuclear plant fueled by enriched uranium, but it can be operated on the spent fuel waste from present plants. It utilizes about 60% of the original uranium fuel compared with about 20% now utilized. It is claimed that by 1990 enough uranium fuel waste will have accumulated to keep breeder reactors operating for 100 to 400 years.

Whatever the outcome of the present struggle, two points seem clear: First, little or no breeder reactor power will be commercially produced in the United States in the 1980s, and second, the breeder will not assume the important place expected of it, even in the 1990s. However, its advantages are so great that the safety and cost problems are likely to be overcome by the time U.S. natural gas and oil production falls off sharply. Until that day, coal will be king; but all other energy sources must be developed as much as possible if a calamitous energy shortage well before the year 2000 is to be avoided.

4. *Solar Energy*

The vast extent of solar energy can be appreciated from the fact that only about one-billionth of the sun's energy is received by the earth, and yet this fraction represents a rate of supply at least 10,000 times as great as the highest rate of the demand for energy by humans and their machines. It seems then that technology should be able to harness this input to relieve any shortages of fuel that could possibly occur. Such speculation is attractive, and it should certainly impel us to investigate thoroughly both the methods for impounding the current sunshine and the methods by which this same flow of energy was stored over the millenia to give us our fuel supplies. Research in both these directions has been meager, however, because the stored product, fossil fuel, has been plentiful so far, and it has been cheap, so we have not really needed to develop means for utilizing solar radiation as an energy source.

Since our twentieth-century civilization developed its large energy demands, an increasing number of enthusiasts have contributed to pioneering efforts in the utilization of solar energy. Direct-collection methods, direct conversion to electric energy by photovoltaic devices, and indirect solar power from winds, tides, waterfalls, and ocean temperature differentials, as well as solar energy storage by photosynthesis, have attracted serious attention, study, and experiment. No doubt we should pay more attention to the enthusiasts who ardently declare that "the age of solar energy has arrived."[24] The two major solar energy contributions to date, for direct conversion, are the commercial development of solar water heaters for domestic use, and the wide utilization in space programs of photoelectric conversion by solar cells. In addition, there are several experimental installations, both in the United States and in other countries. Developments to date furnish a good indication of the near future by emphasizing that there is already a limited use of solar heat for *low-temperature* applications, but there is as yet little prospect of large-scale solar *power.* The solar cell is excellent

[24] See W. E. Morrow, Jr., "Solar Energy, Its Time Is Near," *Technology Review,* **76** (2) (December 1973) 14 pp; also "Solar Energy Is Here But It's Not Yet Utopia," *Fortune,* **XCIII** (2) (February 1976) 13 pp.

for small power demands in remote or inaccessible locations, but as yet it is too costly for other commercial uses.

Solar energy enthusiasts keep reminding us that the hot desert areas of the earth's surface total about 800 million square miles and that it would take only 0.4% of this area, covered with solar collectors, to supply all the present energy needs of the world, even at 10% efficiency. But most deserts are in the wrong place, and there is no great promise today that the capital costs of large-scale solar power plants can be brought down to an economical range. Perhaps some large-scale use for heat, such as in seawater desalination, can in the distant future be accomplished by solar energy. Even if and when solar electric power production at a large desert site becomes feasible, such a project must await the development of economical long-distance electric power transmission. There is as yet no prospect of solving the latter problem in the next 25 years by any method that is economically feasible. At present the longest economical transmission is about 400 miles.

It may be true, as solar power advocates say, that if only one thirty-third of 1% of the area now devoted to farming in the United States were covered with solar collectors, the energy so collected would be the heat equivalent of all electric power used in this country. But this is not really such a small area. For instance, it would be almost equal to all the area of all the roofs in the country (covered with solar collectors). The comparison is also a poor one on technical grounds, because the sun's heat cannot be converted to electric power by any means that is now economical, and the land area in question is needed for food production anyway.

Over 20% of all energy demands in the United States are for heating and cooling in residences and smaller buildings, and most of this energy is now supplied in the valuable form of natural gas, refined oil, or electric power. Solar residence heating, both for comfort and for domestic hot water, has been used on a small scale in many parts of the United States for years. Up to 1974, however, there were less than a hundred good plants in operation, not counting small units for domestic hot water and others for heating swimming pools.

If the present costs for materials, construction, and maintenance of a small residence heating system are charged on the basis of a 25-year life, the unit cost of solar heat is not too much above that for electric resistance heating for the residence.[25] But it takes a solar heating enthusiast to prefer the solar heating system, because of the many convenience factors that favor conventional methods. If summer air condi-

[25] See *ASHRAE Handbook,* Applications Volume, for technical details of present systems.

tioning can be added to the solar plant, the cost is much reduced because the fixed charges can then be spread over year-round loads. As fuel prices increase, solar heating becomes more desirable, especially for those locations where only a small amount of supplementary heat is needed for poor weather conditions. Commercial development of the all-year solar plant is within view and will become attractive as economic conditions change; but there are many problems, as indicated in the following.

Flat-glass solar collectors are used almost exclusively for present heating installations, with thermal storage and a supplementary heat source for adverse weather conditions. The maximum solar heat irradiated and the amount transmitted through each square foot of horizontal flat glass are indicated in Table 15.[26] Of about 440 Btu per square foot maximum solar radiation intensity just outside the earth's atmosphere (the solar constant), not much over 300 Btu per square foot reaches the earth's surface, and of this maximum about 10% is dissipated in transmission through a pane of ordinary window glass normal to the sun. It should be noted from Table 15 that the total transmitted through 1 square foot of a horizontal, flat collector for an entire *day* of full sun is not much over 1000 Btu. This total can be increased by keeping the collector surface always perpendicular to the sun's rays.

The approximate total solar heat theoretically available per 24-hour day is also given in Table 15 for two latitude locations and for each of the seasonal quarters. Any value for the heat transmitted through the collector glass, however, selected from this table, must be reduced to account for the heat lost by imperfect insulation and for night radiation. All values in Table 15 apply only to full-sun conditions, with no clouds, rain, fog, or pollution—and clean glass.

In a great many places, especially in and near industrial locations, and in winter, clouds and pollution reduce greatly the solar heat available; and since the heating plant must be able to carry the load continuously and its heat-storage capacity is limited, it is necessary to provide a supplementary heat source in the form of an auxiliary heating plant.

Summarizing, the ability of each unit of area of a collector to utilize the heat represented by the solar constant (440 Btu per hour) is *reduced* by: (1) the heat absorption of the normal, clear atmospheric envelope, (2) the clouds, vapor, fog, or pollution anywhere within the path between the sun and the collector, (3) the angle of incidence at the collector surface, that is, the deviation of the collector surface from perpendicular to the

[26] Table 15. Solar Heat Available in U.S. Cities: Maximum Heat Quantities per Square Foot.

sun's rays. These losses of solar heat are largely uncontrollable, with the result that, during most of the hours of the day, and especially in northern and industrial locations for most days during the heating season, less than 25% of the 440 Btu per hour per square foot available from the sun outside the earth's atmosphere can possibly be collected at ground level (as an average during daylight hours for a solar heating system with a fixed collector area). This fact, plus the high capital cost of the system, the high maintenance (cleaning), and the need for a supplementary heat source when cloudy conditions prevail, make the solar system expensive compared with any fuel-fired system, except in very sunny and mild locations.

Research in the use of solar energy for residence heating, cooling, and water heating has recently been modestly supported by the NSF (see their reports). Many of the projects have been carried out at universities with industry cooperation. The economic phases have not been neglected, and tables of comparative costs for various cities have been carefully estimated.[27]

In early 1975 the interest in solar energy started to spread at a phenomenal rate and is still accelerating. The doubling and tripling of energy costs and the prospect of their further escalation was a great stimulus and quickly generated dozens of energy symposia, publications of all kinds, and a sudden expansion of both government and commercial efforts in the field. In a great variety of solar heating systems both technical improvements and lower costs followed each other rapidly. Most installations have been individually designed, but many standard components are now commercially available. As the hundreds of installations increase to thousands, quantity production begins to appear, and costs are rapidly being reduced, especially as competition becomes a factor.

For house heating the age-old "passive" methods are being revived, meaning large south and west window areas or massive sunny-side walls which store large amounts of heat. For indoor comfort, these methods leave much to be desired. Window areas produce blinding glare and sudden heating in limited areas only. Heat stored in massive walls cannot be turned off as the day becomes warm. People accustomed to the automatic control of indoor room conditions within a degree or two with gas or electric heat are dissatisfied with passive solar heat which is almost uncontrollable.

Solar space-heating systems that separate the three functions, solar

---

[27] For a short summary see, A. Weinstein and C. S. Chen, "Feasibility of Solar Heating and Cooling of Buildings," *Professional Engineer* **44** (2), 28–32 (February 1974).

heat collection, heat storage, and heat distribution, are necessarily expensive. But they can be adapted to excellent automatic control and have the added advantage of possible combination with conventional equipment to provide year-round comfort conditions.

In most locations in the United States the periods of overcast or dark weather during the cold season are long enough so that solar heat must be supplemented by heat from conventional equipment that has nearly full-load capacity. If the solar heating system has a very large heat-storage component, however, a much smaller supplementary system using fuel or electricity can be used, with thermostatic control activated by the stored-heat temperature. The most attractive possibilities are combinations with summer air conditioning. Such combined systems are especially economical in operating cost if they can use off-peak electric power at lower rates, or if they use a heat pump with its thermal advantage. Another possibility is absorption-type refrigeration for summer air conditioning, in which case concentrating-type solar collectors would be needed, employing either mirrors or plastic lenses to produce a water temperature in the range well above 200°F. Many combined systems are being installed.

One rather extensive development in Illinois uses a very large water tank for heat storage for both space heating and indirect heating of service water. The space heating is by warm air. By using mass-produced components, including vacuum-insulated solar collector tubes, it is possible to combine the cost advantages of solar heat, the thermal advantage of the heat pump, and off-peak discounts on electric power. No fuel is burned in the system.

The enthusiastic activity of solar heating advocates has already resulted in a lowering of first costs of solar plants, but if the additional cost of a solar plant over a conventional plant is to be amortized in a reasonable time by fuel savings, flat plate collectors must be available at considerably less than $10 per square foot. For a collector with good design and ample insulation this has not yet been attained. Here the limitations are expensive materials and fabrication methods, but the latter costs are fast being reduced as the demand volume increases.

The accelerating growth of solar heating may make it a substantial economical competitor in new heating installations well before 1980, first for swimming pools and domestic hot water, and then for residential and commercial structures.

Solar power plants for the large-scale generation of electricity is a matter entirely different from solar heating. The circulating fluid in a solar collector for a steam power plant must provide a heat source above 500°F instead of delivering water in the range 100 to 200°F used in building heating. The higher temperature calls for focused-type solar collectors, usually parabolic cylinders, and much more insulation. Recent studies estimate construction costs for a solar plant at more than five times the cost of a conventional fuel-fired plant. This cost is not prohibitive, since it will be reduced with further development and the cost of fuels is increasing while solar energy is free. It does indicate, however, that several years of research will be required to place the large solar system plant in a competitive position.

The photovoltaic cell, extensively used in space programs, is the alternative to solar steam power. Such cells furnished the energy supply for the Skylab space station for 171 days, even after one wing of cells was lost. At about $300 per watt of capacity the first cost of this power source was too high but, at one-tenth as much, photovoltaic systems are now commercially available for small, isolated installations. Although many experimental solar steam plant installations have been established, especially in Europe, the photovoltaic method is so much less cumbersome for electricity generation that several large U.S. corporations are now working diligently to reduce their cost. With government encouragement, these projects have the ultimate objective of making large solar-cell installations competitive with fuel-fired and nuclear plants, that is, in the range of first cost $500 to $1000 per kilowatt. This is beginning to look distinctly attainable. Among the accomplishments to date are the Mobil-Tyco ribbon process for making solar-cell silicon, and the cadmium sulfide substitute for silicon, in a project in which Shell Oil is cooperating.

Both the direct solar steam and the photovoltaic power methods require large collector areas and full sun. These requirements could be met by stations located in space satellites, which proponents claim would be feasible, given adequate financing. Estimates are as low as $2000 per kilowatt for several large space stations with their accompanying microwave receivers, each several miles in diameter. These costs seem rather unlikely, but the idea is being seriously promoted.

Almost every recent popular article extolling solar energy carries some prediction of the extent of its future use. A favorite forecast is 10% of all U.S. energy by the year 2000, and 25% by the year 2025. These optimistic predictions are seldom accepted. Up to 1973 most of the authoritative predictions for the energy future of the United States did not even mention solar energy as a significant economical source within the next 25 years. Though problems remain, both technical and economic, the challenge is so great that the pioneering should by all means be extended and given much wider support.

Among the *indirect* solar energy systems that have been proposed, the "energy plantation," or forest, is promising. The proposal is to grow fuel for conventional steam-electric plants located adjacent to managed forests. The land requirement would be substantial if the individual plant capacities of today's fuel-fired plants are to be approached, and little research and development have been done because of the scale required. Even though the size of such a plant is limited, the proposal is attractive in areas of adequate rainfall, for example, in the southeastern United States, because no irreplaceable resources are used and pollution problems are minimal. The possibilities of large-scale energy-crop farming, focusing on sugar-producing crops, are being investigated for ERDA by Battelle Laboratories (Columbus, Ohio). Sugarcane, sugar beets, and sorghum are high-yield annual crops, and energy production could be combined with sugar manufacture, and possibly also with the burning of municipal wastes. Only a minor source of energy could be provided in this way, since land could not be spared from food production for the millions of acres required for a large "energy plant."

Other solar sources, namely, wind, tide, and ocean thermal differences, can make only minor contributions to energy shortages in the next 25 years. Each of these may be attractive for small plants in special locations, or for dispersed usage, but their potential is very small compared with that of nuclear energy, for example, or even with that of direct solar energy. (See discussions, next section.)

5. *Geothermal, Wind, and Other Sources*

Geothermal energy is in a class by itself, both as regards ease of utilization and the vastness of the potential supply. Hot springs, geysers, and steam vents from the earth always make a vivid impression, and the amount of heat in molten lava from volcanoes is terrifying. For good reasons, however, geothermal sources make only a slight contribution to the world energy supply. Development should be accelerated, however, even though early results may be very modest.

Italy and New Zealand have long been users of hot water and steam from the earth for electric power generation, and in Iceland most of the winter heating is from natural hot springs. Japan makes good use of its extensive hot springs for baths, as do many other countries, including our own. There are small geothermal power plants in the USSR and in Mexico, and the recent 400-megawatt installation at Geysers Field, north of San Francisco, suddenly made this country the second largest in output of electric power from geothermal sources. World total power generation from geothermal heat is less than 2000 megawatts, however, and rapid expansion is not to be expected.

Few geothermal sources are near centers of population or industry; and because fossil fuels have been cheap and are easily transported and stored, besides being adaptable to practically all uses of heat and power, there has been little incentive to develop the fixed and inconvenient sources of geothermal energy. The conspicuous supplies of hot water and steam are at low temperature and pressure and often contain sediment and impurities such as dissolved salts and gases, including methane and hydrogen sulfide.

However, in view of the high temperature of the earth's interior, large reservoirs of heat must be available; and at rifts in the crust, as well as in volcanoes, they are known to exist near the surface. There are several known areas where the rock mantle is thin or faulted, including western United States, the East Pacific, East Africa, Italy, Russia, Iceland, and so on. Heat is available from pressurized hot water or brine, steam, molten lava, or basaltic rock, and from great fields of hot, solid rock at depths that can be reached by available drilling methods. As fossil fuels become more expensive, it will be economically feasible to develop such geothermal sources, especially after some of the new ideas for transmitting electric power over distances of 1000 miles or more have been developed into practical methods.

If and when the price of electric energy increases into the 5- to 10-cent range per kilowatthour (1970 dollars), geothermal reserves on the scale of hundreds of thousands of megawatts will become attractive. It is fairly certain that power generation of this magnitude, once it has been developed, could be continued for centuries. The vast costs of these new technologies make such possibilities of little concern to the present generation, however, because they are not now economically justified, even though they seem technically feasible.

Developments of geothermal power over the next 25 years will doubtless be modest, continuing along the same lines already used, involving winter heating, industrial and biological process heating, and power generation by low-pressure steam turbines or by thermal cycles using fluids such as Freon or isobutane. In locations where large reservoirs of hot water are known to exist at pressures well above 1000 pounds per square inch, deep hot-water wells can be developed, say, to 20,000 feet in depth. Experimental systems for tapping the vast hot-rock heat storage are already being considered, and several plans have been developed.

One advantage of geothermal energy is its low and localized environmental impact. Much of the water and steam is pure enough to be used with only mechanical separation of sediment, no chemical treatment, and final surface runoff. Dual-well systems are practical, a deep supply well and a shallow waste well. Where brines or noxious gases (mainly hydrogen sulfide) are involved, chemical recovery methods would usually be profitable in a large plant. Most plants will be distant from populated areas, will be inconspicuous, and will not spread environmental pollution.[28]

Wind power, although it has been widely used for at least 1000 years, is not highly attractive as a continuous power source because of the vagaries of the winds. Single windmills also have small capacities, and aggregations of large windmills are viewed as an environmental nuisance, unsightly and even dangerous. It now seems likely, however, that windmills will again be produced in quantity as they were at the turn of the century, when 100,000 or more were in use in America alone.

Especially since the NSF-NASA workshop on wind energy conversion in 1973,[29] both experiments and major design projects have flourished in Oregon, California, Oklahoma, Montana, Vermont, Massachusetts, and elsewhere. There is similar activity in Canada and in Europe.

There have been many serious proposals to utilize wind power on a large scale. These have ranged from plans calling for hundreds of windmills in New England, or in Oregon, to a proposal for 300,000 of them on the Central Plains. As for large windmills, the two best known experiments were the 1250-kilowatt Putnam-Smith wind-powered electric generating station that operated intermittently during World War II (1941–1945) near Rutland, Vermont, and the 200-kilowatt plant built in 1957 at Gedser, Denmark. The two-blade airfoil propeller of the Putnam windmill measured 175 feet across and weighed 16 tons, while the Danish windmill had three 40-foot blades. Several other machines of 100- to 900-kilowatt size have been built and operated in France, Germany, Hungary, and elsewhere. The large windmills have had the benefit of the best modern engineering design and construction, but they have been essentially one-of-a-kind experimental machines, hence the costs were high. In each case the wind data available were inadequate for the best design. For these large units the economics of competition with conventional fuel-fired plants has not yet been demonstrated, but extensive estimates for large systems indicate that they could well become competitive as fuel prices increase.

Even if large-scale wind power might be economically feasible, other difficulties will limit its use. Computed total wind energy available in an

[28] For extensive discussion and references see P. Kruger and C. Otte, *Geothermal Energy*, Stanford University Press, Stanford, Calif., 1973, 360 pp.
[29] For the NSF-NASA workshop proceedings, see NTIS publication PB 231–341, 650 AV, Government Printing Office, 1974.

entire region such as New England or the Southwest is an awesome figure, comparable with all the power presently used in the United States; but the number of wind wheels that could be installed would utilize only a miniscule part of this. Moreover, each installation would deliver year-round less than 10% of the wind energy it received. Some winds would be too strong, and others too weak; hence energy storage must be involved in any system that requires continuous service. In competition with a single 1000-megawatt steam station, large windmills, say in the range of 500 to 2000 kilowatts each, would fill the countryside with hundreds of high towers. The proposed massive "Great Plains network" envisioned 15,000 towers, each 850 feet high and each carrying twenty 50-foot-diameter propellers. While such a system was estimated to have more than one-half the capacity of all the U.S. electric plants, what would the customers do when the wind died down? Environmentalists already object to the few small towers that support high-voltage electric lines.

It is not likely that electrical utility wind power will be developed to carry loads of any significance to assist in the present energy shortage, but small windmill units are another matter. Thousands of such generating units are already in operation in many countries. The thousands in the United States preceded the great drive for rural electrification by the Rural Electrification Administration (REA) and utilities. But even now a program for the mass production of fully automatic windmill electric plants in a range of sizes, say, up to 25 or 50 kilowatts, could serve a great many users. The cost of the windmill plant and batteries would in many cases be less than the cost of running utility pole lines to their locations. Such locations are now largely served by gasoline- or oil-engine plants of low overall efficiency.

Present subsidized research and development programs emphasize wind power plants of less than 100-kilowatt size, and these programs should be expanded and continued. Although the total contribution cannot even approach 1% of the energy load within the next generation, such installations will save scarce fuels, can be economically feasible, and should be promoted. Today the installed capacity of electric generating plants is only of the order of 500 million kilowatts, so 1% would be 5 million kilowatts. If windmill plants averaged 50 kilowatts each, 100,000 of them would be required to attain this 1%.

6. Miscellaneous Energy Converters

Several developments in the energy field do not qualify as energy sources but rather as improvements in energy conversion. After further development they may be expected to take their places in the eventual

economic solution of the energy crisis. Among these devices are mag-
netohydrodynamic (MHD) conversion systems, fuel cells, new electric
batteries, and thermoelectric and photoelectric generation.

MHD or plasma generators convert fuel combustion energy directly
into electricity by means of an ultra-high-speed plasma tunnel with
powerful external magnets. They thus substitute a continuous-flow device for the combined functions of a complex boiler-turbine-generator.
They use ultrahigh temperatures; and when installed as "topping units"
for a conventional turbine plant, the result is a high overall efficiency.
They are still in an experimental stage, and miniature as compared with
their expected size in a conventional plant of high-megawatt capacity.
The higher efficiency of the dual unit may reduce the overall fuel consumption by 20%, with the plasma entering the generator at about
5000°F and exhausting to the conventional plant at 4000°F or less. The
problem of containing gases at these high temperatures, in large-scale
equipment, has not yet been solved, however.

Fuel cells are electric batteries with a fuel input. They thus bypass
the extensive combustion engine–generator system and accomplish
direct conversion from the chemical energy in the fuel to electric energy
at the cell terminals. Hydrogen has many advantages as the fuel, and the
hydrogen-oxygen fuel cell is already highly developed. More easily
available fuels such as natural gas, methanol, and gasoline are attractive for the user and may eventually be substituted for hydrogen. Aside
from the space program, the main applications have been for small
isolated plants; but there is promise of commercializing a unit of
moderate capacity which will provide electric power and heat, mainly for
residences using natural gas fuel. Such a unit is already in limited use.

Electric batteries of new types, both primary and secondary, are
already used for many smaller applications. Potential uses are large, but
costs are high. For example, a new electric storage battery of high
current density, light weight, and reasonable price would be ideal for
propelling a small car and could revolutionize the automobile industry;
but such a battery is not yet available at a reasonable cost.

Thermoelectric generators, that is, power thermocouples, and *photoelectric generators* or solar cells are more or less in a single class. Both
use expensive materials to produce very small electric power output at
low efficiency (10% range). The thermocouple can be reversed and used
for refrigeration. Solar cells are widely used in space vehicles, but they
are as yet too expensive for commercial application. Both these devices
are commercially applied only for small-output units located in remote or
inaccessible places.

Development and application of all new energy systems and devices
should be encouraged and supported. Energy supply has become criti-

cal. Every new source and every improvement in the utilization of present fuel sources has potential for alleviating the energy crisis, especially if its operation does not add pollution to the environment.

E. PROJECT INDEPENDENCE, OR SELF-SUFFICIENCY BY 1985

President Nixon's early proposal for Project Independence, to make the United States self-sufficient in energy supplies by 1980, provided a real stimulus for discussion and planning. Gradually the many complexities are becoming better understood, but it will be some time before all aspects of energy independence are appreciated. One approach that recognizes both computer analysis and judgmental forecasts by specialists was well presented in mid-1974 by a policy-study group from the Energy Laboratory at MIT.[30] Here an attempt was made to evaluate the effect of energy prices on supply, using a range of $7 to $11 for crude oil. Attainable rates of production from all present energy sources by 1980 were carefully estimated (Table 16).[31] The effects of some of the methods of regulation, such as price controls, tariffs, quotas, and federal leasing, were also examined. Attention was called to the fact that a stockpile of imported oil would provide a hedge against import disruption.

The general conclusion of this study was that energy independence by 1980 could be purchased only at a very high cost in economic resources, environmental degradation, and possibly even income distribution. Minimum attention was paid in this study to the future beyond 1980, to sources of capital, or to techniques for initiating the program or modifying it as it progressed.

Since the convenience and comfort of almost every resident in the United States depends to a startling extent on continuing supplies of petroleum and natural gas, the proposal that we will again attain the self-sufficiency we enjoyed years ago, from our own resources, has tremendous appeal. Now the earlier enthusiasm for Project Independence has been overshadowed in the daily news, but nevertheless a great deal of study and planning toward the president's proposed goal of self-sufficiency has actually been done, and a twenty-volume government report issued. The entire program calls for work on every one of the methods for increasing our energy production, with early starts on each

---

[30] "Energy Self-sufficiency: An Economic Evaluation," *Technology Review*, **76** (6) 23–58 (May 1974).
[31] Table 16. Energy Projections for 1980: Energy Self-Sufficiency—Prospects and Results.

method and acceleration of those found most promising for quantity production soon.

One obvious way to assist toward self-sufficiency in energy is to improve utilization. Here there exists a tremendous opportunity, and this opportunity is gradually being embraced. A favorite legislative method is to set arbitrary goals such as a reduction of a million gallons a day in oil imports, or a fixed miles-per-gallon figure for new cars. The great waste of energy at present affords both personal and technical opportunities to improve the efficiency of fuel use. The Ford Foundation energy report, "A Time to Choose—America's Energy Future" (343 pp., 1974), gave so much emphasis to the conserving of energy that it immediately drew sharp criticism from the oil companies, from many others in the fuel industries, and from several Congressmen from fuel-producing states.

One technical example of an opportunity for fuel savings is in the original design for a large complex such as a shopping center. The overall costs for power, lighting, heating, and refrigeration can often be much reduced if a central plant is designed to provide all services for all tenants instead of each tenant being served independently for each of these needs.

For most energy services the economic emphasis has heretofore usually been on first cost, be it an industrial, commercial, or home installation, or even a new car. This might be expected, because both manufacturing and construction are so heavily sales-oriented. Low first cost often means·high operating cost. "Life-cycle costing" is finally being more widely considered by purchasers, especially for large installations. Congress has even discussed requirements for life-cycle costing for all federal agency facilities. This would call for an in-depth study for each construction project, estimating not only first costs but also predicting as accurately as possible operating costs for the expected useful life of the structure or plant. The total life-cycle costs would then be compared in deciding on construction contracts. This is not new, but it is not yet widely practiced. However, as energy prices increase and more and more kinds of energy service are being required by users or occupants, it becomes an important means for conserving fuel.

A similar example is the total cost of personal transportation. Here everyone is concerned, but there are many false ideas about economy. Vehicles could and should be made much more efficient, and this is finally getting some emphasis. Slower speeds and adequate maintenance save gasoline, and so do car pools. But the substitution of public transportation can be self-defeating, especially if the total transit and waiting time of each passenger is taken into account. In a specific 1974 study, the federal Department of Transportation (DOT) considered a major switch from cars and trucks to mass transportation. It concluded that the shift would take 10 years to complete, would require an investment of $6.2 billion, and would finally save 1.8% in energy use. This saving would certainly not pay the interest on the investment, not to mention lost time and travel inconvenience.

In the long term, Project Independence has two interrelated objectives: (1) Develop entirely new sources of energy such as the breeder reactor and solar or geothermal energy. (2) Change the relative use positions of the common fuels, especially by generating electric power from coal and nuclear reactors rather than from oil and natural gas.

Depression in business plus large unemployment and serious inflation have unfortunately created a very poor atmosphere in which to launch a great national research program. Public apprehensiveness and distrust of the government had been running high when President Ford, in October 1975, announced a comprehensive proposal for an "energy independence authority." His plan was comparable with the great space programs of earlier years and envisioned a total effort that might approach $100 billion over 10 years, primarily to develop nonpetroleum energy supplies. The President's first announcement was made in a short address to the AFL-CIO, and this did not allow time for details, but he did outline a method for financing, essentially by means of loans and loan guarantees. Research and development would be spread largely to private companies, and the federal government would make or guarantee the loans they required to support their work. With so few details revealed about the program itself the method of financing immediately drew criticism, largely on the basis that it might increase inflation and exert an upward pressure on interest rates. Besides, it was letting the government make all the decisions as to what was to be done and who would obtain the loans, it was interfering with the free market, and the method of repaying the loans was unclear. These criticisms, by Congress, by the public, and even by government planners and business people, eclipsed any consideration of the project itself. When the President sent his plan to Congress, they were merely repeated.

As a basis for evaluating the desired changes in the energy industries in terms of their present status and past experience, Table 17[32] gives statistics on output, employment, and finances for the years 1950–1972 (see also Table 14). The total employment in these industries is not large, being just over 1 million as compared with over 11 million in durable goods manufacturing and about 3.5 million in contract construc-

---

[32] Table 17. The U.S. Energy-Supply Industries, 1950–1972: Output, Employment and Finances.

tion (1970). The utilities in particular are capital-intensive. In 1970 the plant investment per employee in electric utilities was about $320,000, and in gas utilities $240,000. Operating revenue in electric utilities the same year was over $75,000 per employee, and in gas utilities over $100,000 per employee.

Complete domestic self-sufficiency and cessation of oil imports may not be the only method for protecting the United States against disruption of essential energy supplies or excessive oil prices. A middle-ground solution can probably be attained and is now being advocated. In any event, the problem of balance of payments must be taken into account. Either of the extremes, maximum imports or zero imports, introduces serious problems, among which are the difficulties of enforced reductions in oil consumption. Extremely high capital requirements may well make Project Independence impossible to attain fully, especially in view of the recent performance of the capital markets. The physical near impossibility of meeting construction deadlines by 1980, or even by 1985, should also be mentioned.

Every energy-saving proposal must be fully analyzed for long-term costs. When large capital investment is necessary for an energy-saving project, a study is indicated to determine whether an investment of the same amount in the development of new energy resources may be more advantageous. In any case, all investments of this kind are long-term projects, one reason being that slow starts are made because of pressures from various special interests. Massive oil imports are, in the meantime, piling up other problems for the future, while Project Independence is lagging. As a method for fuel savings, high prices for fuels have one decided advantage in that they automatically encourage marginal sources and substitutes that would never pay otherwise.

F.   REDUCING THE U.S. ENERGY USE

There is every prospect that the United States will be forced by circumstances to reduce its energy demands, which have been growing since 1950 at rates of 4 to 8% per year (Table 10). If total present consumption were actually to be curtailed, it would force drastic reductions in our energy-based economy which serves a growing population. The preferred objective then is to reduce the yearly *rates* of growth in energy consumption. A shrinking economy is, to most people, unthinkable, but slower growth might be accommodated.

Focusing specifically on the 1970–2000 period, certain projections of future growth have already been quoted (Section B). Another forecast, .

from government sources, is presented in Table 18.[33] Energy quantities in this table are expressed in uniform units (Btu $\times$ 10^{12}), and growth rates are given. It is noted that total energy consumption is assumed to increase at a rate of slightly more than 3% per year, or a growth rate per capita of something over 2% per year, to the year 2000. Solar and geothermal sources are not mentioned. These predictions are modest in comparison with some others, but in view of recent developments even these seem too optimistic. There is little prospect that supplemental supplies of oil in the year 2000 will equal domestic production, even if maximum imports are allowed and synthetic oil development is intensively promoted. It is even less likely that nuclear energy will increase more than sevenfold in the 20 years preceding the end of the century, a growth rate of more than 10% per year.

Many aspects of the U.S. energy problem have been carefully studied during the past 5 years. Earlier predictions relative to the balance of the century were often just optimistic extrapolations of the prosperous decades of 1950–1970. Following the OPEC oil embargo a dozen or more additional investigations were launched by fuel users, the fuel supply industry, the automotive industry, the federal government, the military, and private agencies such as universities and educational foundations. Later reports then predicted slower rates of growth. Recession experience and the immediate prospects for more of the same have subdued much of the optimism. Knowledgeable authorities are becoming convinced that energy use per capita cannot continue to grow at past rates and that the limited supplies will soon call for great efforts toward energy conservation. The realities of slower growth, and the prospect of actual reductions in some areas of energy use, are forcing wider attention. It seems probable that, in the near future, high prices, shortages, regulations, both federal, state, and local, and voluntary organization to reduce energy consumption, will all be necessary. The federal leadership shows preference for control by market forces with emphasis on higher prices and education of the public in more efficient use of costly fuels. State public utility commissions have allowed modest price increases, but accompanying campaigns for economical use have been largely absent.

Voluntary organization to eliminate waste and the unnecessary use of energy can readily be encouraged if regulation is strict and prices are high, but first the public must be convinced of the realities. Ever-increasing fuel imports shift more and more U.S. employment into goods and services that must be exported to pay for the imported fuel. Petro-

[33] Table 18. One Forecast of U.S. Energy Consumption to the Year 2000, Energy Sources and Uses (Btu $\times$ 10^{-15}).

leum imports have been exceeding one-third of our total consumption and, if the 50% mark is approached, we will become highly vulnerable to disruptions in our energy economy by outsiders, and to prices that include large gifts to the OPEC.

In conformity with the slow growth future that now seems to be dictated by the limits of supply of critical materials, and by the limited capacities of the human environments offered by spaceship earth, this author has selected lower growth percentages, and these are shown in Table 19.[34] This new recommended assumption of a *slow growth* pattern includes energy growth rates as follows: yearly growth of total U.S. energy consumption, 2.5%, doubling the supply in 28 years; yearly growth of the electricity supply by utilities, 4.5%, doubling in 16 years and supplied largely by coal and nuclear reactors; and yearly growth of the petroleum supply, 2.0% doubling in 35 years. It is recognized that rates of growth will fluctuate somewhat, but it is assumed for convenience that these will be the averages over the 5- or 10-year periods. No consistent increase in the rates of growth is projected.

Our recommended projections of energy growth in Table 19 may be compared with some of the earlier projections already quoted herein:

1.   For the *total energy supply*, the NSF study of 1971 (footnote 16) predicted 2.81% average annual growth; the studies of Table 18 predicted slightly over 3%; the NASA report of 1973 about 4% to 1985; the *slow growth* rate we selected for Table 19 is 2.5%.

2.   For *electric energy* generated the NSF study predicted 5.46% annual growth; many authorities are even now advocating 7% growth; Table 19 projects 4.5% growth.

3.   For *fuels*, Table 18 predicts about 2.5% annual growth for coal and oil but a negligible increase in the supply of natural gas. Table 19 projects a high rate of growth for coal supply, 4.5% annually, but only a 2% annual increase for oil and gas.

In a very real sense the spectacular U.S. economic growth during 1950–1973 could not have been accomplished without corresponding growth in the use of energy. Table 19 therefore emphasizes the crux of the entire question of future economic growth. We now find that large yearly increases in fuel consumption just cannot be maintained. U.S. production of natural gas and oil is slowing down, and proved reserves are small. This, among other things, forces a slow growth economy. Early

[34] Table 19. U.S. Growth to the Year 2000—Preferred Assumptions: Historical Growth and Realistic Future Slow Growth.

adjustment to this slower rate of growth after the recession will help to prevent major future crises which would force a shutting off of electrical and gas supplies to many customers and also the rationing of petroleum fuels.

G.   CONCLUSIONS ON THE U.S. ENERGY POSITION

The United States will continue to be short of petroleum and natural gas. Perhaps we shall never again have as much as we think we need. Since the reserves of coal are ample, every effort consistent with environmental limitations should be made to increase the production and use of coal, including its use for synthetic gas and oil. Programs to develop all domestic sources of oil and natural gas should be intensified, and the use of the several additional sources of energy should be expanded, with most of the emphasis on nuclear and solar energy. Efforts to counter the shortages of petroleum and natural gas should be directed first toward expanding energy sources that promise large blocks of energy output for conventional usage. This means that expanding the supply and the direct use of coal is of first priority, followed by the extraction of synthetic gas and oil from coal and shale. Geothermal, wind, and ocean sources do not qualify for quantity output, but their development should continue. Solar energy for generating electricity, and also nuclear fusion and direct-conversion methods for electrical generation such as the fuel cell, must likewise be regarded as sources for the distant future, since they call for large expenditures and lengthy research and development programs before mass-production results can be expected.

There are relatively few alternatives for the immediate intensive program. Certainly major efforts and large amounts of capital must be expended on increasing the supplies of coal, petroleum, and natural gas by methods already known. Coal, rather than oil or natural gas, should be used for electric power generation. Next in priority are expanding nuclear power usage and developing the breeder reactor. There will be delays in attaining quantity production of oil and gas from coal and shale, because the processes are only at the pilot-plant stage, and both technical and environmental problems are formidable; but these projects should be intensively promoted. Solar energy for heating, in small-capacity units, commands wide interest. It is ripe for commercialization, and its adoption should probably be accelerated by subsidy.

There have been many loose claims to the effect that one-fourth to one-third of our fuel consumption could be saved by eliminating waste and improving efficiency. This might be true if waste were defined as "all unessential consumption" but a tremendous effort would be required, with the elimination of many conveniences we have come to regard as necessities. A few effective measures have already been adopted, the 55 miles-per-hour speed limit, for example. But saving a third of the fuel now used to heat a building, or a like fraction of the gasoline used to drive a car, calls for much redesign and replacement. Most buildings and most cars are not new. They were bought when fuel was cheap and perform accordingly. New buildings and new cars that use fuel more efficiently, without pollution, will be more expensive. Industry uses a third of the fuel supply (Tables 11 and 18) and has similar problems in fuel saving.

It takes time to carry out the necessary engineering work that must precede more energy-efficient cars, and much additional time to make changes. Such time spans can be compressed by legal regulations. An early example of this was the enactment of legal limits for exhaust emissions from new 1968 cars and the subsequent extensions of this act. Here is another example of the interdependence of the four crises involving energy, environment, population, and food. The rules about car exhaust emission were enacted to control pollution, but the first methods that were applied to accomplish this greatly reduced gasoline economy, and following the oil embargo this became intolerable. The struggle of automotive manufacturers to deal with the two problems in an economical manner is well illustrated in the report, "Should We Have a New Engine," issued by the Jet Propulsion Laboratory in 1975 (JPL SP 43-17, 660 pp., commissioned by the Ford Motor Company).

Food supply and population growth are a part of the problem. Oil prices and engine economy are of first importance to all farmers, and high oil prices soon affected food prices. Almost 20 million persons were added to the U.S. population from the time Congress enacted the exhaust emission regulations to the end of 1976. The population growth continues, calling for more food and more fuel, and adding to the pollution.

Congress, the people's representative, has been slow in accepting the recommendations of its expert advisers and of the president in calling for legislation that sets up mandatory fuel conservation. The 1975 Energy Policy and Conservation Act was largely concerned with voluntary conservation programs. It set up some government assistance, largely by grants, to encourage energy conservation planning by the states. Proposals from a state to the federal government were invited by this act. In these proposals, mandatory thermal efficiency standards would be listed. For both new and renovated buildings, standards for insulation, lighting, and controls would be included.

Efficiency labeling of appliances such as air conditioners, refrigerators, and freezers, was made mandatory by the act. It also charged the president with initiating a ten-year plan for energy conservation in buildings occupied by federal agencies and set up another plan for "energy efficiency improvement targets" for industry. In the latter plan the aim was to persuade the large industrial energy users to arrive at a "maximum feasible improvement" as a target for 1980. This means a mandatory reporting to the FEA of plans and action to reduce consumption of oil and gas fuels. These measures to reduce waste and raise efficiency in energy conversion and use have little opposition, except as to their potential for government dictating to management, and the imposition of much paperwork. Resistance to these could easily block the real achievements, however. Component performance and prescriptive standards for energy conservation in buildings were dropped from the 1975 bill, but these were presumed to follow in a short time. When such prescriptions are to cover a major building complex, a controversial situation is created. Presumably such performance requirements were to be left to state codes, but if and when they become part of federal requirements, even those for federal buildings, they begin to resemble a national building code. A rigid national code could stifle important innovation and hold up progress for years.

The closest approach to a national building design proposal has been the move to specify the adoption of "ASHRAE Standard 90," developed by the American Society of Heating, Refrigerating, and Air Conditioning Engineers. An extensive study of the 1975 edition of this standard, ASHRAE 90-75, was funded by the FEA and completed by Arthur D. Little, Inc. This standard has been adopted by several states and is under consideration by several others. Among the important findings indicated in this report were that larger buildings originally designed to conform to ASHRAE 90-75 would show reductions of 40 to 60% in annual energy consumption compared with conventional building designs. It was also estimated that general adoption of ASHRAE 90-75 could result in such annual market shifts as $179 million more for thermal insulation, $175 million less for lighting fixtures, and $185 million less for heating and air conditioning equipment, plus a considerable increase in the time and effort spent on engineering design.

The controversial aspects of such a standard as ASHRAE 90-75 would emerge, especially in the vast federal building programs and their implementation. Another area of controversy would center on state versus federal authority. Energy conservation aspects might fade in comparison with economic aspects, for example, fixed investment charges versus operating costs. The life-cycle costing equation could be juggled by bidders on large contracts.

The conflicting interests of the various parties involved will no doubt be gradually resolved, but the entire concept of mandatory performance standards creates apprehension about high initial costs and long-term bureaucratic enforcement. There are many who point out that current energy costs are still low in comparison with other products and services. It may therefore be better to proceed as many other countries have, impose very high fuel use taxes and leave it to private initiative to make all the improvements that will eventually save fuel, such as much smaller cars and more efficient processes and plants.

A government move that will expedite energy developments in the future was the establishment in 1975 of the ERDA and the NRC. The new ERDA was made responsible for the development of all energy sources, taking over these efforts from several agencies, including NASA, the former AEC, and several bureaus in the DOI. But centralized authority for the energy conservation program remains an elusive goal. ERDA's Office of Energy Conservation has broad programs, but FEA, FTC, and the Department of Commerce have also been given assignments.

While progress is being made on all fronts toward a better U.S. energy position, this progress is slow. Perhaps this is the democratic method, and as the action gathers momentum much more will be accomplished. Barring emergencies such as another oil embargo, an extremely cold winter, or a major war, U.S. energy supplies can hold out for the few years it takes to organize more economical use and develop substitute sources, but only if a large program gets under way soon. Unfortunately, the majority of citizens are not well informed, so they are still unconvinced, and this seems to include Congress. Doubtless, small goads to vigorous action will soon appear, such as blackouts, industrial closings, long filling-station lines, school closings, and cold buildings due to a shortage of fuel. We will then hasten to improve our energy position. Unfortunately, our very slow start postpones and even endangers major success.

The increase in domestic energy output in the next 25 years can be only modest as compared with its growth in the recent past, summarized in Table 17. This table shows a striking contrast in the rates of increase in use of the four common kinds of energy, oil, gas, coal, and electricity. Coal production has hardly increased at all. Oil production doubled in the 23 years from 1950 to 1973 but could not keep up with demand, so major importation was necessary. The output of natural gas, which climbed to almost four times the 1950 figure, showed slower growth than that of electric power output which increased to almost six times the 1950 figure.

As to the future of U.S. oil and gas production, perhaps it is a good thing that price controls have held down industry expansions somewhat.

Domestic reserves promise to last 15 to 25 years, depending on rates of consumption (Table 13), but resumption of more active prospecting for new oil and gas fields and increased supplies from other sources (coal, shale, nuclear, solar) could more than double that period.

One of the major deterrents to an adequate U.S. energy supply is public attitude and consumer agitation. This is well known to industry and government leaders, but it is seldom mentioned. Poorly informed and unthinking customers organize protests against price increases that are necessary and inevitable in the adjustment to scarcities. Given adequate information, the American public and its legislative representatives usually make wise choices; but ignorance of the facts can result in highly detrimental moves. A main reason for this book is to provide a concise but sufficient source of information on conditions today and on the experience of the past in the energy field (among others). From such information the citizen is enabled to form intelligent opinions about proposals for the future.

Projections of slower growth such as those in Table 19 are necessary as a basis for examination of the adjustments for the next 25 years. Such projections afford some guidance for planning within the limitations of the capital, manpower, and environmental protection that must be taken into account, while preserving our system of private enterprise and representative government for a U.S. population which will grow by more than 50 million from 1970 to 2000 (see Chapters 6 through 8).

This slow growth pattern will be resisted by the entire economic system and by our "sky is the limit" philosophy of growth. But the adjustment to slow growth is a great challenge; and the pattern has many advantages, even though the higher costs of all energy will demand intensive practices in fuel economy.

Whatever the extent of energy demands in the near future, the expansion of domestic supplies will be limited. The limiting factors of capital and manpower are examined in Chapter 7, and some of the estimates of future requirements of the energy industries are given (see Tables 18, 19, and 38).

Chapter Four

Food Technology and Food Supplies

A. FEEDING MORE AND MORE PEOPLE

A hundred years ago the world had ample surplus land, including large fertile areas not being used for farming. Fishing was small and local and could hardly be called an industry. The age-old struggle for enough food seemed to be ending. In America, great areas of excellent farmland were just being brought under cultivation, with ample help from capable immigrants from the Old World. The railroad and the steamship were solving the problems of food transport.

Fifty years ago world population passed the two billion mark, but there was little concern about the adequacy of supplies of food from land and sea. In the United States and Europe especially, people were then almost unaware of any threat to their ever-increasing supply and variety of foods.

When the total world population passed three billion in the 1950s, however, the predictions by Malthus (1798) that the population would outrun the food supply began to be reconsidered; but to most people in the United States a food shortage seemed highly improbable. Had not mechanization permitted great improvements in farming methods and allowed large-scale operations? Pesticides and chemical fertilizers ensured good crops, weather permitting. Improved storage took care of the lean years, and government subsidies canceled many of the serious risks.

Four billion people on the earth has now been attained. It suddenly becomes apparent that most of the four billion are poor, and even hungry. It is fully expected that world's population will reach six billion within about 25 years, and then what about the food supply? In the United States where well over 40% of our diet consists of protein-rich foods of animal origin—meat, fish, eggs, and dairy products—it seems unbelievable that over half the world's population have no such choice of food, that 90% or more of their food comes directly from field and garden crops, and that malnutrition and protein deficiency are widespread.

World peace now depends on our recognition that such great disparities in wealth and food supply do exist, and in turn on the efforts of the wealthier countries to help in improving the situation. When world population growth adds a number equivalent to the entire U.S. population every 2 or 3 years, and doubles the world's population in two generations or less, it is time to examine the prospects of feeding these future billions. How can food production be doubled in the next 25 years, and then doubled again by our grandchildren in the following 25 years? What caused this explosion in the demand for food, and how can future demand be controlled and met? The answers are not simple.

In the first place, the population explosion has been real. Now the world totals of tillable land are seriously limited, as are also water supplies for human and plant use, nitrogen supplies for plant growth, feed for animals, and consequently high-protein foods. Even today, the world's livestock consume almost four times the quantity of food consumed by humans, and large additional amounts go to pets and wild animals. (Pets are not common in China; they cannot spare the food.)

Consider the possibilities of expanding farming to twice and then four times its present production within the lives of children today. Most of the best land is already under cultivation, with a world average of about 1 acre per capita. It has been estimated that the world's tilled land could be doubled, but not much more. In any event, capital costs will be high. Some of the additional land must be cleared of timber, scrub, heavy growth, and rocks. Some of the land must be drained. Some of the land is remote from roads, transportation, and communication. Some of the land is marginal as regards weather, moisture supply, and topography, so that special contour farming or irrigation must be established. These things require extra labor and cost money.

Careful surveys indicate that total world food consumption today, when animals are included, is equivalent to feeding almost 20 billion people. The greatest problem is in Asia, which has about 56% of the world's population and only one-third of the arable land. Farms there are actually "miniplots," having been subdivided for generations. In India, Bangladesh, and Indonesia, most of the farms are less than 5 acres, and animals eat much more food than is used by the human population. The diet available in India, Indonesia, and several other Asian countries averages less than 2000 calories per day, and less than 10% of it is of animal origin, so the total protein content is low. Food production and diet in several African and Latin American countries are almost as poor as in Asia.

At this late date modern agricultural methods are certainly known to all of these nations. What are the reasons for low food production and malnutrition in so much of the world, when most of the population in these countries are farmers? Many of the reasons for these conditions are obvious, and some have already been mentioned. Inadequate education, large families, crowded farmland, exhausted soil, small-scale and primitive methods, poor transportation, large animal populations, and general poverty make improvement very difficult. Many of these countries are in tropical or semitropical regions which experience alternate seasons of dry, scorching heat and torrential, soil-leaching rain. Attitudes of lethargy and fatalism are generated, and there is then little interest in the hard work and further sacrifice required to introduce new methods and improvements, or to move to better locations. Such changes also require capital, of which there is almost none available.

Most of the world's food has always come from cereals and other edible crops. Additional protein is obtained from animals and from fish and other sea life. A latent possibility is that foods might be synthesized chemically. The possibilities of startling improvements in food production, and of new kinds of foods, are examined in the next section. These may offer promise for the future, but their introduction will be slow, and it is worthwhile to examine food production today and to estimate the prospects of an adequate, continuing food supply based on present technology and resources. First let us look at the quantities involved and examine the prospects in the light of recent experience.

There are actually two problems. Because many of the world's people are even now undernourished, the food supply must first catch up to provide enough for a healthful diet for everyone, and then steadily increase and double in 30 years or less because of increasing world population.

World population grows by almost a quarter million babies a day, and the children come first. Children need more protein than adults, and a prime responsibility of this generation is to see that they have it.

Currently the growing world calls for a food supply that increases by at least 25 million tons of grain every year. One of the reasons for soaring food prices is that this required addition is almost twice as large as it was in 1950, and it will continue to grow. In most developing countries the demands for food are rising much faster than domestic food production, so that large grain imports are needed year after year. It is estimated that 84% of the six billion or more world population in the year 2000 will live in Asia, Africa, and Latin America.

The FAO has predicted a sixfold increase in importations of grain as early as 1985. The United States and Canada are the world's largest food exporters. The exportation of over 72% of the U.S. wheat crop and 60% of the oil seed crop in 1972–1973 represented large increases from the early 1960s. The U.S. share accounted for 58% of the world's exports of oil seeds and cake in 1972–1973. Exports of rice remained at about 60% of U.S. production for the 1963–1973 decade.[35] These gains improve our precarious balance of payments in the face of the oil crisis. In view of the world food situation, they are likely to persist in spite of their unwelcome effects on U.S. food prices at home. Our great preference for meat and dairy products might even be affected. Certainly additional farm acreage and higher productivity per acre will be required.

The U.S. population will increase perhaps 50 million before the year 2000 (Table 4). One question that will soon arise is how much of the added grain production to that date is needed for home consumption. If the rest of the world were fed today at the same standard as the United States, it would call for world total food production at least six times as great as it is now. The United States and Canada use annually a ton of grain per capita, as compared with an average of 400 pounds in vast areas of the poor countries. Can we expect to maintain this difference in the face of world food shortages? Actually the relative production of the poor countries will probably decrease, because the price of fertilizers and fuel for trucks and tractors more than tripled during the early 1970s and is still increasing. For families who are already spending more than three-fourths of their income for food, the only possible response to tripled prices is less food.

Two methods for reducing this great inequity immediately present themselves, cheaper food and more food. Cheaper food means fewer pounds of grain used per million calories of human nourishment, without a sacrifice in the protein content of the diet. The ready answer to more food for humans is less food for animals.

A million calories of corn or wheat, eaten as the common cereal foods on our tables, represents the farm product of about 1 acre. A million calories of beef requires the output of almost 10 acres, plus 5 acres or more for hay land and pasture. In these same terms of land usage, potatoes and sugar are less expensive than cereals, and poultry and

---

[35] Above statistics are from the "International Economic Report of the President," submitted to Congress, February 1974. (Superintendent of Documents, No. 4115-00055.)

eggs half as expensive as beef. Other costs rise with the acreage too. Production of a pound of beef might use 3500 gallons of water, a pound loaf of bread one-tenth as much. U.S. consumption of beef and chicken per capita doubled from 1950 to 1970.

This preference for meat is expensive in terms of the amount of vegetable protein consumed. A pound of dry, raw soybeans contains almost twice the protein of a pound of beef, and it costs very much less to buy. Peanuts and most other beans also have more protein per pound than beef. Oatmeal contains only slightly less protein than lamb.

A prolonged U.S. energy shortage would make it extremely difficult for our farmers to increase production as much as is needed. Today the ratio of raw energy input per food calorie in grain output is at least 10 to 1, much higher than earlier in the century. In order to double our grain production, much more than double the energy now used by our farmers will be required (see next section).

What about the capital costs of massive increases in U.S. farm output? Here again the necessity for an overall systems approach is clear. Capital demands for a dozen other purposes will be in competition (see Chapter 7). Additional plants must be built to produce the added fertilizer. Large investments by farmers for more farm machinery will be essential, and energy production must be increased to supply the additional machines. Since everyone else wants more energy too, the capital demands of the energy industries will be very high. Added investment in agricultural research is needed, because some of the added crop production must come from better seeds and advances in farm methods. Additional crop land will be required, calling for investments in land clearing, drainage and flood control, or dams and irrigation. Most of the good cropland is already in use. Throughout, the environment must be protected, adding further costs. It is apparent that the assignment to double food production is a typical systems project.

Market forces alone cannot be expected to control capital allocations, so some independent agency, probably government-related, should make systems studies and recommend priorities in the competition.

Returning to the prospect of more food for humans by feeding less to animals, we note first that less than one-fifth of the world's grain production goes directly to human food. Most of the rest is fed to meat animals, draft animals, poultry, and dairy cows. One-half of the world's agricultural land is worked by draft animals. A substantial quantity of grain becomes food for wild animals, pets, birds, rodents, and insects, or is lost by spoilage. From the U.S. grain crop, over 85% of corn, barley, oats, and sorghum, and 90% of nonexport soybeans become animal food.

Millions of tons of milk and thousands of tons of fish (meal) are fed to meat animals and poultry.

One much-quoted estimate is that U.S. grain crops must quadruple in the last 30 years of the century. Even if this were successful, there would be no assurance that a much larger percentage of the grain would then be available for human food. An intensive effort to increase this percentage would make a difference, however.

Wider cultivation and greater human direct use of crop-type high-protein foods is most promising. Many poor areas already substitute high-protein beans and peas for meat. These legumes at the same time enrich the soil with nitrogen. Perhaps the price structure will gradually enlarge the use of vegetable protein and fats. Butter and lard are already much less widely used, meat extenders (from soybeans) are appearing on the market, and peanuts and beans are becoming more favored. The recent trends away from cereal foods might be reversed by only a little persuasion.

These methods, together with the development of new high-protein foods, and a greater interest on the part of governments, plus worldwide cooperation focusing on the food problem, will go far toward reducing malnutrition and warding off starvation in the third world, at least buying time for a generation or two. The eventual solution must be ZPG.

B. THE LIMITS OF FOOD PRODUCTION FROM LAND AND SEA

There is widespread malnutrition in the world. A serious examination of the limits of future food production is urgent. Are there prospects that the food supply will reach its limits in the near future, even while the total world population is increasing faster than ever? In 1925 the world's population was twice as large as it was 100 years earlier. In 1975 the world's population had doubled in just 45 years, instead of 100 (Table 3). The developing countries are growing over twice as fast as the developed ones, and soon will have three times the population. The annual U.S. population growth was 1% or less during 1930–1970, except for a bulge around 1950 (Table 4).

Because our own rate of growth has decreased and our food production keeps increasing, it is difficult for the U.S. citizen to be concerned about the limits of food production. Food surplus has been our problem. While our population has increased about one-third since 1950, the output of our nine principal food crops has doubled, while the number of farms has been cut in half (Table 8). Food productivity per farmer thus quadrupled in 25 years. Productivity per acre also doubled, as farm acreage under cultivation decreased slightly.

Stories from the persistent advocates of ZPG, plus the many statistics collected by the UN and others, are beginning to convince us. There *is* a food supply problem, and it promises to become much worse. The question of how much the yearly world food production can be increased, and when, is already urgent, but the answers depend greatly on what people are willing to do. When farmers have large crops of grain, cattle, and dairy products, and prices go down, they may shoot calves, pour milk down the sewer, and demand more grain exports immediately. If generations of farmers in India and Bangladesh have been using simple plows drawn by oxen or water buffaloes, they resist change. They have no money or credit to buy better equipment anyway. Famines are a cruel way to educate the world to the growing food problem. More food from present sources and new sources and new foods are the evident needs.

1. Food from the Sea

Enthusiasts remind us that the lowest level in the chain of ocean food, plant plankton and seaweed, represents, in the North Atlantic alone, a total quantity thousands of times greater than the grain and rice crops of all countries combined. This potential food source is not easy to utilize for humans, but efforts in this direction should continue.

Perhaps the greatest opportunity for early increases in human food from the sea would be accomplished by greater concentration on fish that feed directly on plant and zooplankton—herring, sardines, salmon, whitefish, and so on. Carnivorous fish which are several steps beyond in the producing chain—tuna, pike, haddock, and halibut—represent an input-output ratio of at least 10,000. In other words, 1 pound of these fish is the result of at least a five-step chain which started with 10,000 pounds or more of plankton. Even compared with beef production, this is a very wasteful way to produce protein by the biological process. Doubtless some day the marine biologist and the ocean chemical engineer will develop a process that is much more efficient, with a product that is just as palatable as tuna or pike; but for the present we much prefer to let nature do the processing even if we must substitute herring and salmon for tuna and haddock in order to stretch the supply. Great quantities of fish meal are now used for animal food, especially for chickens; and here is a possibility of more direct use of these fish by diverting them to human food.

Most fishing today is near the coasts, in relatively shallow waters above the continental shelves, and the possibilities of real deep-sea fishing are largely unknown. We do have experience with another method for extending the fishing areas. We have some fish hatcheries, and we look forward to using them on a large scale to service "fish ranches." Again, this is not a likely development before the twenty-first century, although it might be if a crash program were undertaken.

Starting with present UN activities, and after the 1974 Law of the Sea Conference, it is hoped that international agreements will set up a program to exploit the several possibilities for rapid expansion of human food supplies from the sea. But cooperative agreements and plans come very slowly. In any event, greater attention to fish food resources may serve to restore the rapid growth of the fish catch. In the early 1970s the fish catch declined, after a fairly steady growth for 20 years during which time the world fish catch, as reported by the FAO, increased from about 20 million tons to over 70 million tons per year. At present fish is less than 2% of all human food, but about 10% of the protein.

The United States has recently been the world's largest importer of fish, and we have taken more than one-half of all the tuna. Even so, our per capita fish consumption is small (less than 12 pounds annually), and two-thirds of our available ocean protein goes into animal foods. It would not be strange if the protein-poor nations should demand some of this for human food, especially since 65% or more is lost when it is fed to animals.

On the whole, the promise of food production from the seas appears to be a modest one as far as the near future is concerned. We appreciate and need enthusiasts who will develop new foods from the sea, by hydroponics, mariculture, and other specialties, even though their expectations of quantity are not realized. At present we must conclude that increases much above 2% of all human food, coming from the sea, can be expected to come only very slowly, if at all. But this 2% is still very important.

2. Food from the Land

Agricultural production in the developed countries has been increasing so fast that any possibility of a world food shortage seems to us entirely unreal. In food crops, the developed countries can keep way ahead of their population growth, while the developing countries are unable to keep up with theirs. From our viewpoint, practically all underdeveloped countries cling to primitive methods of agriculture; and although most of the working population are engaged in that field, unemployment and malnutrition both continue to increase. World food production patterns per capita since 1960 are shown in Table 20.[36]

The United States made several fortunate and wise early moves which contributed to its agricultural development. These included its early immigration policies, the Homestead Act of 1862, the Morrill Act of the same year, and the Hatch Act of 1887. These not only expanded agriculture into our vast areas of good land, but they furthered agricultural art and science by great programs of education, research, development, and assistance to farmers.

Even before the turn of the century we were one of the best-fed nations on earth, but it is not within the scope of this book to outline the marvelous farm developments that accomplished this. The marvels have continued, however, and in addition to our vast resource of fertile land and the capabilities of our immigrant farmers and their descendants, no small reason for the progress is our attention to agricultural education and research, with its parallel development of the farm equipment and food processing industries. When we think of great research accomplishments, we are now inclined to connect them with the military and space programs; but we overlook the fact that prior to World War II the largest single area of federal research programs was agriculture (about one-third of the yearly total).

We have had a long struggle with surplus farm commodities. As late as the early 1960s, our carryover stocks were 40 million tons, or almost a ton per U.S. family. Gradually, as our agricultural exports grew, these stocks came to be valued as "cushion" against crop failures, here and elsewhere. Now these surpluses have dwindled to an almost negligible size, while our agricultural exports have increased to an all-time high. In 1973 our farm product exports were valued at almost $13 billion and comprised 70% of all our commodity exports. This places us in a pivotal position as regards world food supplies. With a negligible surplus or reserve, suppose we should have a U.S. crop failure and have produced only enough to feed ourselves—or even less?

In the United States we do not *need* to produce more food than our recent crop averages, but we could readily do so. Our complex agricultural program, with its soil bank and its price supports to keep farmers' incomes on a parity with the incomes of others, has kept farm production down. Even so, the United States and Canada have become the most

---

[36] Table 20. Food Production per Capita, 1961–1972, Index Numbers Based on 1963 = 100.

important sources of food grains to compensate for poor crops in other areas such as the USSR, Asia, or northern Africa.

We are returning to the position where emphasis must be placed on *increasing* food production rather than limiting it. We welcome the crop exports to balance our growing imports, and our farmers appreciate the added income. Assistance to the developing nations calls for large amounts of food, and there are many other nations (including our own), where some of the people are hungry or malnourished. With world population still growing nearly 2% per year, the United States is called on to play a major part in the corresponding growth of world food supplies; and this means that we should keep a stored surplus as crop insurance. There is little prospect that many of the underdeveloped nations will be able to meet the requirement of tripling their food production within the next generation; and unless the United States and Canada do at least this, starvation will become widespread.

Most of the *unused land* in the United States that might possibly produce good crops is idle because of lack of water, although there are a few areas of swamps or flooding and minor areas of depleted soil. Both irrigation and drainage are costly, especially for the locations remaining that call for such treatment. It is assumed that the U.S. Soil Bank Program is about terminated and will not be renewed. High food prices are probably here to stay, and these will encourage new projects. But there have been some disappointing experiences involving U.S. reclamation projects, and broad ecological studies must precede any new moves.

Bringing new land into cultivation is largely a problem of balancing costs and return, but there are strict limits to the amount of land available; and if population and food demand continue to double every generation or two, those limits will be reached rather soon. The FAO reports that North and Central America utilize for agriculture (including pastures and forests) about 25% of their total land area and about 40% of the land that appears to be potentially productive.

Large areas that would otherwise be agriculturally productive are in use for cities, towns, highways, railroads, and so on, so it would be impossible to double the total area used for agriculture. There are deserts and swamps that could be made productive, but at a very high cost. It is true that cropland could be increased at the expense of pasture land and forests, since less than one-fourth of the total agricultural land is now planted to major crops. Even with this increase it would be difficult to attain a doubling of the cropland without a major change in our style of life, including reductions in meat and dairy foods and in forest products, and greater limitations on recreation areas.

Worldwide, the picture is not greatly different. Agriculture, including forests, uses about 30% of the total land area, and 45% of the land that is potentially productive for agriculture (FAO data). Doubling the productive agricultural land would be an outside limit, and very costly; but some of the pasture lands and forests could be diverted to the cultivation of major crops. There are of course great variations among different countries. In Europe, the percentage of agricultural land in use for major crops is already almost twice as great as in the United States, while in Australia and South America the relative amount of cropland is extremely small.

Bringing more land into cultivation is especially costly in densely populated countries. But even if world cropland could be doubled (which it cannot), the yields from much of the new land would be lower, and the demand for fertilizer high.

With the necessity of tripling food production in the short period 1970–2000, it is evident that the old reliable method of increasing the cropland area will not suffice. And what about our great-grandchildren *after* the year 2000?

The subject of fertilizers is complex and controversial. The rich nations are the large users. The hungry nations, where two-thirds of the world's people have little more than one-half the world's arable land, use only about 10% of the chemical fertilizers, and much of this on export crops. Farmers must use fertilizers properly; and although the FAO and others conduct many fertilizer demonstrations, the fact remains that, in an underdeveloped country with many thousands of small farm plots, farmers cannot acquire either the technique or the cash for their extensive use.

Should great fertilizer plants be built in the underdeveloped countries? A well-planned program of building such plants might be better foreign aid than sending food, but there are difficulties. The raw materials, phosphate, potash, and lime, plus energy for the nitrogen-fixation process, are scarce in many countries. (It takes about 50,000 Btu to make a pound of nitrogen fertilizer.) Transportation for distribution of the fertilizer is difficult and expensive.

Efficient use of fertilizer calls for *better seeds*, and more water for larger crops. There are countless examples of high-yield varieties that do well in the American Midwest but are not resistant to the conditions and the pests in India or some African country. Furthermore, the necessary moisture may not be available or, if it is, it comes in downpours, all at once. Even where large-scale farming is possible, the recent inflated prices of machinery, tractor fuel, and fertilizer constitute an impossible combination for a poor country.

In the United States farmers are always striving for higher yields; and with land-bank payments eliminated, they will cultivate more land. Both intensive and extensive operations call for more fertilizer. This means not only capital for a great expansion of fertilizer factories, but also large projects for the transportation and application of the fertilizer, not to mention increased energy consumption and the water required. Present quantities of fertilizer transported in the United States constitute the equivalent of an annual U.S. delivery of over 10 million fully loaded 10-ton trucks. Commercial fertilizers do not restore the soil completely, so soil depletion is still a possibility. If current animal wastes could be effectively used, it would be a great insurance against soil depletion.

The "green revolution" envisioned for poor countries, through the use of high-yield varieties of wheat, rice, and so on, cannot solve their food problems. Crop increases can be attained that keep up with the increases in local population for one generation. But what about the following generations? It has also been demonstrated that, when great areas are planted with one variety of seed and a destructive pest arrives, the results are much more devastating than if thousands of farmers plant their own seed, representing thereby many sources and varieties, some of them more resistant to new pests.

It is true that in several locations two-crop yields are a possibility for a single year by substituting fast-maturing varieties for the usual native seeds. Very frequently the weather does not cooperate, especially in the tropics where a season of destructive drought may be followed by storming downpours which wash away any crop that has survived the dry heat. Another problem, almost as insoluble, is the lifelong habits of the workers. The leisurely way of life that has persisted for generations in a tropical country has a gentle way of defeating an energetic, two-crop program.

The general conclusion about expanding the food supply, tripling it by the year 2000, is that it can certainly be done if all available methods are intensively utilized. Rather large investments will be required, local shortages will be experienced, and the relative prices of foods will almost surely increase. Moreover, there will be great pressure on the United States and Canada to increase their food crops rapidly, to help fill the growing needs of the developing countries. There is serious question, however, as to how these massive exports are to be financed (see Chapter 7, Section A).

It is important to note that most of the methods for increasing food production are one-shot remedies for shortages. They cannot be repeated. Even a seemingly permanent improvement in the use of marginal land, such as a dam and irrigation system, is relatively short-lived.

Throughout the world, including the United States, hundreds of such water conservation systems have been abandoned. The causes of early failure range from great silt collections behind the dam, or excessive salt deposits on the fields from a highly mineralized water supply, to complete failure of the original water intake source. When any new water-conserving irrigation system is planned, ecological studies should be included, and its rather short life recognized in the overall scheme.

Once an area utilizes its surplus land, irrigates or drains where necessary, uses better seeds and ample fertilizer, and raises two crops a year where practicable, that is the end; and if demand still increases, there is little prospect of meeting it. Control of population growth is the only solution; and since the United States and many other countries have a built-in growth potential beyond the year 2000, there will be intense pressure to develop entirely new sources of food (see next section).

The final method, *storage*, or carryover of product, is as old as the Egyptian pharaohs and the ancient Chinese emperors. Of course, it does not expand the food supply, but it can ward off shortages and famine in lean years.

Oddly enough, the great crop storage programs in the United States after our dustbowl experiences were designed to protect farmers' incomes and prices rather than to protect consumers in case of devastating crop failure. In general, the programs were successful, although not entirely in terms of their stated objectives. The amounts of sheer waste will never be known. In later years they did furnish incentives for several aid programs, domestic and foreign, including school lunch programs, food stamps, and "food for peace" exports.

It is now evident that surplus crop storage was not a foolhardy remedy. With the whole world looking to North America for a food supply, it even becomes good insurance to have a reserve in case of major crop failures anywhere.

The extreme limits of future rates of production of today's foods are highly dependent on international cooperation in all phases, including storage and distribution. Such extensive cooperation could result in good times for farmers everywhere and in expanding opportunities for individuals throughout the system. Employment should be increasing in agricultural research, and in design and construction of fertilizer plants, dams and irrigation systems, and new tools and machinery. Additions to storage and distribution systems will be needed. Broad technology transfer and user education are required, and this will call for many professional specialists.

South America, with one-third of the planet's river waters, should make more use of them in agriculture. In Africa, the almost 3 million square miles dominated by the tsetse fly and other pests must finally be conquered and agriculture expanded accordingly. In India, Pakistan, Bangladesh, and other crowded countries in which many farms are only the size of large city lots, newer farming methods should be introduced, small-scale tools and machines developed and shared, and millions taught how to use them.

In oil-producing areas enough flared gas is wasted to make all the nitrogen fertilizer the world could possibly use. This is an opportunity for the developed nations, because the technology is complex, the plants expensive, and the transport problems awesome. So is the financing.

The hunger belt contains two-thirds of the world's population, four-fifths of the babies, but only one-fifth of the food production. There are many ways to improve this imbalance in addition to those already mentioned, and in most cases several countries would be involved. Effective cooperation is therefore necessary, and aid from the developed countries must go beyond financial assistance. It must include actual participation to accomplish the technology transfer to the recipients, with demonstrations and education of the millions of small farmers.

Present-day food supplies can probably be tripled, but in the meantime the search for new foods must be diligent. These are inspiring challenges, and there is no room for pessimism.

C. THE PROMISE OF NEW FOODS

It is not safe to oversimplify the possibilities of "new foods from chemistry," even though many of the essential components of human food, such as vitamins, amino acids, and so on, have been synthesized. Any widespread use of synthetic products that are nourishing and healthful is certainly an accomplishment for the rather distant future, not to mention the economic problems involved. Nor are humans easy to convince as to substitutes for familiar foods. Experiences in introducing wheat in rice-diet areas, or of gaining acceptance for meat-substitute products are recent warnings of this. The most promising prospects for chemical synthesis of food probably do not lie in the complete synthesis of the carbon-containing complexes of natural foods. Rather than start with carbon and its simple compounds from petroleum, coal, or papermill waste, a conversion of available vegetable materials that are undigestible by humans, to a digestible product, offers promise, especially since the vegetable materials used as raw material are plentiful, cellulose, for example. Biological assists to the chemist, from bacteria, fungi, and so on, are of no small importance in the picture of mass production of new foods.

Although the expansion of present natural processes of food production from land and sea gives ample promise for meeting the needs of the next generation (see previous section), it might be argued that we should expand the use of synthetic nutrients as rapidly as possible to supplement natural foods. A good start has already been made in the widespread vitamin enrichment of flour and cereal foods, and the use of the many additives on the GRAS (generally recognized as safe) list of the Food and Drug Administration. A next step might be the use of synthetic amino acids as a food additive or supplement for protein enrichment. At least this has already gained success in animal feeding (e.g., poultry).

New foods and corrective or supplementing mixtures are never easy to introduce, even though they may be of natural origin. Wide acceptance is still more difficult to attain. A combination that is palatable in Latin America may be completely rejected in Japan or the East Indies. Americans may eventually accept certain additives in hamburgers or hot dogs, while somewhere else they would be better received if used in soup or in fried meal cakes. Combined foods are not necessarily inexpensive, and government aid programs or subsidies may be necessary to attain the ultimate aims, such as reducing malnutrition and increasing resistance to disease.

There have been several disappointing attempts by American companies highly experienced in food marketing to introduce enriched-food mixtures in some of the developing countries. Like the rest of us, the customers in those countries are suspicious of most food additives and prefer familiar flavors and consistencies. Another problem in the introduction of new foods in the less developed countries is the wide dispersion of the population on farms and in small villages, and the lack of good communication and transportation. It is much easier to sell food supplements to the developed nations, just as it was easier to persuade them to use chemical fertilizers and to plant improved hybrids.

The providing of additional *quantities* of common foods for the present and future billions of people is much less formidable than meeting the demand for *quality* and low price in this additional supply. The success of any large plan for introducing new processed or manufactured foods is critically dependent on both low cost and high acceptance. An additional difficulty is that the raw materials and the market are usually far apart. These several problems are minimized when the new food is a high-quality, enriching additive which can be produced locally from a material normally associated with some staple food.

There are many opportunities for meeting these criteria, and processing technologies are already available. In fact, some rather large-scale operations are in existence, such as those associated with the "Meals for Millions" programs. Usually the final product is a high-protein supplement or additive for an otherwise inadequate diet consisting mainly of rice, corn, small grain, tubers, or legumes.

Several food wastes or by-products now discarded or used only as food for animals make promising raw materials for conversion into human food or food-enriching additives. These include skim milk and whey, oil seed meal and cake, fish meal, animal blood, and certain less marketable leftovers of meat, poultry and fish; also by-products of cane and beet sugar manufacture. It is claimed that, from oil seeds alone, as much edible protein could be obtained as all of that now supplied from meat, milk, and eggs. The major oil seeds are cottonseeds, soybeans, peanuts, flaxseed, and sunflower, sesame, and safflower seeds. The residual meal or cake from each of these seeds has one or more constituents that are undesirable in human food. Changing the method of oil extraction may be necessary, or undesirables may be removed by additional processing. Other treatment to change the color, flavor, or texture may also be desirable.

Soybeans are relatively low in cost and have the highest protein content of any common food. Nevertheless they are little used for human food in most advanced countries. The mammoth U.S. crop is largely exported. Fermentation is an ancient method for making food from soybeans. The common raw products from soybeans are soy flour, 40 to 50% protein, and soy concentrate or isolate, 70 to 90% protein.

In a very comprehensive report, "The World Food Problem," published in 1967 for the president's Science Advisory Committee, major attention was paid to the quality as well as the quantity of the prospective world supply of food. A subpanel submitted a report entitled, "Increasing the High Quality Protein," and outlined in detail the complexity, scale, and urgency of this problem. This panel first surveyed existing diets in various regions and considered means for their improvement. Their report presented data on the nutritive value of protein in common foodstuffs, the concentrations of the essential amino acids therein, and the possibilities of fortifying food by means of protein concentrates, preferably with local production of the concentrates. In addition, the report discussed the genetic means for increasing the desired protein in corn and other cereals. The prospects for satisfactory single-cell protein foods, produced by the culture of bacteria or yeasts on hydrocarbon or carbohydrate substrates, were considered, with process data and costs included. For each avenue of approach, the panel emphasized the economic difficulties and also estimated the time that would be required to attain substantial production and acceptance. They indicated that additional basic research and technological development would be required and said it was probably not reasonable to expect such foods to play a significant part in world protein nutrition before the 1980s.

From this report, and from the extensive studies by recognized authorities such as Borgstrom (Michigan State University), it must be concluded that many new protein foods have promise, but they represent a long-term solution and not a quick remedy for the world hunger crisis. The urgency of the problem of food scarcity seems to demand that work be accelerated on all fronts. There are still great possibilities for increasing the quantity of common foods, and for improving food quality in areas where malnutrition and hunger are great and increasing.

In most of the food-shortage areas there is high unemployment. Farming is crude, and farms are small, although production per acre is not necessarily low. Many improved methods and machines used by the large-scale American farmer are not applicable. Improvements should therefore concentrate on the development of small equipment, better tools, and labor-intensive methods. China, the world's most populous nation, has apparently accomplished the transition to self-sufficiency in food. Her methods deserve thorough study.

D.   THE HUNGRY TWO-THIRDS OF HUMANITY

How many people in the world are always hungry? This question should be, rather: How many people must continuously exist on an inadequate diet? We know that there are entire countries, such as India and Ethiopia, where the *average* diet for the entire population is inadequate; but we are also aware that in *every* country, even our own, there are people who do not get adequate food, and often it is not their fault.

A recent UN report contained data that showed thirty-seven of seventy-one less developed countries with inadequate food over a period of several years. In terms of the percentage of the world's population, the ratio is even higher. A U.S. Senate committee reported in 1972 that there were at least 12 million people in our own United States who were undernourished because they were too poor to buy adequate food. This is over 5% of our population, and we are the best-fed country in the world. Moreover, we know individually that there are even others who keep their need a secret; some of them are ill, infirm, or aged; others are children whose mothers are ignorant of what they need.

The inadequacy of food in the poor nations is evident from a study of

Tables 1 and 7. When the *average* diet in a country furnishes less than 2200 calories per day, and protein-rich foods from animal sources are less than 10% of the total, it is certain that very few of the inhabitants have an adequate diet. By UN data, the forty countries in which these minimum amounts are not attained have a population of 1.9 billion, or about one-half of the people alive today. Another 500 million are counted where the average daily diet has less than the 2500 calories designated as the malnutrition boundary for a population with a rather high percentage under age twenty-five, especially if they get less than 20% of their food from animal sources (the U.S. average is over 40%), so that they probably do not get adequate protein. The remaining world population of about 1.4 billion reports an average above the malnutrition level but, judging from the poverty data in Table 1 and elsewhere, at least 10% of them (140 million) are too poor to eat adequately. These very rough estimates add up to the conclusion that at least 2.5 billion of the roughly 4 billion people in the world are not getting adequate food, that is, two-thirds of the human race is malnourished—or worse.

It is near impossible to grasp such vast numbers and to believe them, much less to realize what they mean. Let us get down to specifics. India is the largest country involved in the poverty struggle. With a population of almost 600 million, it has a gross domestic product per year of less than $100 per capita (compared with our $5000). Year after year runaway inflation has made family purchases more difficult. Since oxen are largely used for transport, and food must be shared with them, and with water buffaloes, monkeys, antelopes, elephants, wild hogs, rats, parrots, crows, locusts, and so on, the feeding burden is actually at least three times that of the human population. If India were to eat the entire world catch of fish, its protein diet would still be too low. Hungry mouths are added at the rate of about 15 million a *year*, a new population equal to that of our seven North Central states (Minnesota, Iowa, Missouri, Kansas, Nebraska, South Dakota, and North Dakota). Additional tilled land equal to all the wheat acreage in the United States would need to be added every 4 years, with fertilizer and water in proportion also, just to feed the *added* population. The "green revolution," using high-yield rice and wheat and added fertilizer, has already been more than offset by population growth, partly because water is often in short supply and crop storage facilities inadequate. The refugees from Bangladesh added to the pinch. The indications are that India will have a billion people to feed by the end of the century.

Not only in India is the gross domestic product less than $100 per capita, but also in Indonesia, Bangladesh, Burma, Nepal, Ethiopia, Nigeria, Tanzania, and several other countries. The great food assistance programs of the developed nations and the UN, plus those of hundreds of private agencies, and even the contributions of the rich Arab oil nations, do manage to distribute a major share of the world's food surplus to the hungry. Will these efforts, added to those of the nations themselves, eventually be able to conquer world malnutrition? Unfortunately, the prospects are grim. Malnutrition is on the increase, and conditions will become worse unless the annual population growth in these nations, now 2 to 3.5%, can be reduced, say, to the 1% level (see Table 1). Reducing the world's animal population would help, and some countries have already started such campaigns.

When an agricultural area becomes vastly overpopulated with men and animals, the process becomes irreversible, and there is little hope that the land will ever again meet the feeding load which keeps increasing. Food can be imported, but is there any hope that such importation can be supported in any way except by larger and larger doles from the developed nations?

A current example of this kind is the wide belt across Africa south of the Sahara and Egypt. Almost from coast to coast, except along the Nile and its tributaries, the 50 million people who have been struggling to stay alive by following their age-old patterns of nomadic grazing or primitive agricultural existence are losing the fight. The affected strip starts with Mauritania and Senegal on the west and includes large parts of Mali, Upper Volta, Niger, Chad, the Sudan, and Ethiopia. Smaller areas are similarly affected in Ghana, Nigeria, Cameroon, Kenya, and the Somali Republic. The national governments do not have the resources, human or otherwise, to help to tide these people over the several years of drought. The worst feature is that the gross overpopulation of millions of animals, especially cattle and goats, has so overgrazed the pasture lands, and uprooted and trampled the sparse grasses, that the heavy winds carry away the dry topsoil; and when the torrential rains come they wash away plants and roots. Animals have even eaten the trees. Human populations have been decimated, and herds almost wiped out in spite of the million or more tons of grain sent in for relief.

It is fully expected that the remaining population will again want large families and will build up their herds again as fast as possible, adhering completely to their traditional ways. When the next successive droughts affect the area, the process will be repeated all over again. Similar cycles are experienced in several other malnourished areas, and the gradual process of education for change will take generations. Relief operations will continue, but the capacities of even the best farming areas must prove unequal to the feeding of increasing billions of humans and animals.

E. POLICIES AND ROLES FOR NATIONAL
PARTICIPANTS

Every attempt at international cooperation emphasizes the fierce independence of all individual nations. Seldom has this independence been more effectively demonstrated than it was at the UN World Population Conference at Bucharest in 1974. At the same time that this conference adopted a resolution recognizing that the world food supply is precarious and millions are faced with possible starvation, one after another of the developing nations refused to support any action aimed at the reduction of population growth, the main reason for food shortages in the very countries faced with starvation. The conference finally agreed to declare that population control policy must be integrated into "social and economic development" and that all countries should be encouraged to establish patterns of growth that would eventually stabilize world population at about three times its present size (instead of four times as previously discussed, i.e., at about 12 billion rather than 16 billion). A motto much repeated during the conference was, "The best contraceptive is economic development."

Perhaps it is encouraging that many countries whose representatives champion the free increase in population as a basic right, or for religious or economic reasons, are actually among the two-thirds of the human race that already recognize and practice birth control. Brazil wants to double its population to attain great-power status, but makes contraceptive services available to all. China encourages late marriage, distributes contraceptives, and encourages wide spacing of births, but objects to the publicizing of its low birthrate. India is desperately trying to increase food production and to reduce its rate of population growth, but frequently reminds the West that slower growth assuredly follows economic development.

During the late 1960s the world population increase was about 70 million people per year. Over 60 million of these were added in the less developed countries, that is, the total population added in these countries was over six times that added to the industrialized world. This fact has great significance with respect to food supplies and nutrition.

Food production and food shortages will be very important in the political arena for years to come. The large food producers, especially the United States and Canada, and to a lesser extent Australia, New Zealand, and certain areas in Latin America and the USSR, will be under great pressure to increase their crops for export. The Arab nations are now insisting that the developed-market nations share products, capital, and technical know-how; but at the same time they are collecting for their oil exports at a rate that threatens the stability of every country that is not self-sufficient in petroleum, rich and poor alike.

As long as the pressures are only economic and the United States and Canada are the only countries that can provide large export quantities of food, we in America are in an advantageous position. But as some countries begin to face starvation and others financial disaster, peaceful solutions may be next to impossible. National roles will become critical. Contentions over food and oil supplies loom as one of the great dangers for the next generation. Extreme nationalism in well over a hundred countries increases the difficulties of peace keeping by the UN and by the large nations that try to stabilize the situation.

It is fortunate that large exports of crop foods are advantageous in improving the U.S. balance of payments, but one major disadvantage has arisen. These exports have expanded very rapidly (see Table 21),[37] and the competition drives up food prices here at home. This greatly increases the complexities of farm and consumer regulation by the federal government. A free play of market forces would fail to protect either the American farmer who raises the crops or the American consumer who is apprehensive about high food prices. Promoters and food dealers can easily take out excessive shares for themselves; and if the farmer and the consumer are not protected by federal regulation, large crops will not be planted and export controls may be necessary. Since democratic methods are slow, large instabilities and swings may be expected rather than a smooth and steady process of the expansion of production and export of food. This subjects the democratic system to a serious test, since the impatient consumers are also voters and assume that they have final control of the long-run process. After the manipulations revealed by Watergate, many people distrust government moves, and this too impairs the entire democratic operation.

It becomes evident that the impending food shortages place on every individual a real responsibility to attain some understanding of the world food problem, and the domestic one as well, by learning the facts, taking a firm position, and effectively selling his viewpoint to others.

The United States is in a pivotal position with respect to world food supply, but all the developed nations are called on to help in solving the food problem, warding off starvation, and curing malnutrition. A massive transfer of wealth and know-how from rich to poor nations seems inevitable. Except for emergency food and subsidized foreign trade, most of this transfer should be used for the purpose of increasing the food-

---

[37] Table 21. Exports and Imports of Food: Changes in the Net Foreign Trade of the United States.

producing abilities of the recipient nations themselves. Here are great and challenging opportunities for the younger generation in all affluent nations.

Only seven countries are presently able to grow food crops in a quantity that is consistently above domestic needs, namely, Canada, the United States, Argentina, Australia, New Zealand, Thailand, and Ireland. These countries, with less than 10% of the world's population, face the awesome responsibility of helping the poor countries (mostly in Asia) that cannot produce or buy enough to feed the 35% of the human race they contain. Several smaller nations could well join the food-exporting countries in the near future, for example, several East Asian countries, if they could stop fighting among themselves.

Two great nations could tip the balance, the USSR and China. It is reported that mainland China has been able to end the food shortages that have persisted there for centuries. If their stability continues, they may even join the food exporters.

Russia has not yet solved her agricultural problems; but her grain production has increased, in an erratic fashion. Her 1973 crop was about double that of 1965, but there have been some poor years (1963, 1972, 1974, 1975). A citizen of the USSR has plenty of bread and ample garden produce, and eats more meat than formerly; but about half as much meat as we do. From what we can learn, it appears that over thirty of every hundred workers are farmers in the USSR, compared with 5 in the United States, and the Russian farmer is about one-tenth as productive as his or her counterpart here. But they have many handicaps. Only about 10% of the land is suitable for agriculture, the growing season is short, the weather is unreliable, and most areas are too dry. They cannot grow our two main livestock feeds, corn and soybeans. They have many large farms and some good machinery, but not enough tractor operators, service mechanics, and repair parts. Many residents on collective farms are quite unmotivated, and farm management leaves much to be desired. In spite of all these handicaps, the USSR is the world's largest producer of bread grains, livestock, milk, potatoes, and cabbage.

In the final analysis, the basic problem is "too many people"; and nations cannot much longer disregard this fact.

1. U.S. Roles in the Food Crisis

Before discussing future roles, the present situation and recent trends in exports and imports of food should be well understood.

As shown in Table 21, U.S. exports of grain and animal feed (first four lines, total) doubled between 1960 and 1972 and doubled again in 1 year, 1972–1973. Thus the total commercial exports quadrupled during 1960–1973. (Note: When corrected for inflation by Table 29A, the three ratios were 1.8, 2.1, and 3.8.) It should be observed that the total exports of other grains plus animal food exceeded those of wheat, and the percentage increase of wheat exports was actually lower. As a result of our large imports of coffee, fish, and sugar, and moderate imports of meat, our commercial foreign trade in foods was formerly in near balance, but in 1973 we started a large export surplus. However, over the entire 1960–1973 period there had been sizable additional food exports under the various foreign aid programs, a fair amount of the grain coming from the stored surplus. While our grain imports were negligible as such, meat imports should be interpreted as representing some grain.

The total food crop capacity of U.S. farms now greatly exceeds the country's needs and will doubtless continue to increase. The very fact of this excess capacity introduces many difficulties, but it also adds responsibilities in the setting of future policies. How much of this excess production should be channeled into voluntary public or private aid to needy countries which must import food? In Chapter 5 attention is given to past and future policies and methods of foreign aid. But there are in addition sizable needs for food for the underprivileged in our own country. Several programs for meeting these latter needs have been used, such as food stamps and welfare allowances, but it should be noted that these programs were started while we had surplus grain in storage.

The time has come for every American to accept the fact that there *is* a world food shortage and that it will become a lot worse in the next 10 or 20 years. The United States will have plenty of food, but what has happened throughout history when a small group has plenty to eat while their neighbors, ten times as numerous, do not have enough? America has some problems to face, and they can be summarized under three headings: first, the cost of food; second, problems of the farmer; and third, who gets the food?

Food *costs* are going up, and world food costs will continue to rise, maybe 10% per year for the next 10 years. Americans may suffer less than others, but we must concede that the prices of cereals and meat must include the farmer's costs for fuel, fertilizer, and transportation, and all these continue to increase. There is great pressure for world food production to be tripled in a single generation. We, as the world's largest food exporter, must expect to increase our farm output, and here we must recognize the law of diminishing returns. Each 10% more fertilizer is a little less effective than the previous one. We are using good land, and it costs more to prepare and use marginal land. Irrigation is expensive. Crop failures will be disastrous. These are not matters of money supply or price control, but of resources and human effort. What we need is a wise program for increasing farm output, plus convincing information to the public to obtain everyone's cooperation.

What about the *farmer*? Here in the United States we depend on only about four million farmers, but a good many of them are independent and trying to decide what to do next. Farming involves large commitments every year, and the farmer has little control over output, selling price, or customers. With the entire world depending on exports from the United States, some system of reserve storage must again be undertaken. Hence there are three markets, domestic, export, and storage. Competitive market forces neither allocate the product nor ensure the farmer a reasonable return. A large federal farm program is again needed, especially since the exports must include food aid to needy countries. There will be no more *surplus* grain. Instead, sudden huge demands are likely to appear, which, by their size and timing, may cause a large jump in domestic food prices. Our experience with the 1973 Russian wheat deal must not be repeated. At that time the low price to the Russians convinced wheat farmers that their share was unfairly small. There was a wave of opposition to our role as the foremost contributor to the world's breadbasket and a lessening of the farmer's incentive to raise larger crops.

Incentives for farmers are actually a world problem and, unless these incentives are protected, food shortages will become much worse. An essential government role thus emerges for every major nation, and this role is shared by international organizations.

2. Who Gets the Food?

The FAO expects that by 1985 the *needs* for imported grain will be seven times what they were in 1970. The United States and Canada face an allocation problem. How can we decide who gets the food, even in a normal crop year? What would happen if we have a very poor crop? Who is to be favored among our largest customers, the USSR, Japan, Britain, or India? Allocations must not be decided by how we can make the most money.

Obviously, residents in the countries that raise the food will be in a preferred position, but even here there will be pressures for change. Since World War II, farming has changed greatly in the United States. There are now fewer small, diversified, self-sufficient farms, and many more large specialist farms devoted entirely to the production of animal feed. We export large quantities of feed grains. How long can the heavily meat-eating countries justify a diet that uses four times as much of the world's food resources as would a balanced diet containing much less animal protein?

One point seems clear. As the suffering caused by food shortages increases, the pressures on the largest grain supplier will increase. Population growth is the prime cause of the shortages, but the energy crisis is no small contributor. Many Americans do not yet admit that there is either an energy crisis or a food shortage. Here are pressing assignments in education and communication. Our contributions to the easing of the energy crisis and its attendant environmental difficulties have been severely delayed by disagreement and bickering at top government levels. This will be resolved. The long-term challenge is to proceed with setting up the various means for coping with the interdependent crises in food, energy, environment, and population growth. The means are certainly available.

Chapter Five

Plans and Aid
for the Less
Developed Nations

A. SHARING WITH THE DEVELOPING COUNTRIES

It is now becoming well known that a great many of the world's nations have not yet attained the economic development that ensures the average family an adequate diet, basic education for the children, and living conditions that are better than primitive. In such an underdeveloped or developing country, two-thirds or more of the population are farmers, that is, their vocation, if any, is to provide food. Agriculture is dominant; there is very little industry; and unemployment is high. Agricultural methods are primitive. In most cases the birthrate and the net increase in population are high.

As long as the population increase in such a country continues to be in the range of 2 to 4% per year, the demands of the increasing population make major improvements in living standards next to impossible. A study covering the 1963–1972 period showed sixty-five countries with annual population growth rates in the 2 to 4% range (Table 22),[38] and these comprised over 40% of the world population. Largest of the fast-growing countries were Mexico and Brazil, but countries with annual growth of 3% or above comprised less than 10% of world population.

---

[38] Table 22. Countries with High Population Growth: Rates of Average Annual Growth, 1963–1972.

Aid to developing nations has, over the last 25 years, become an ongoing program. Many countries have participated, but growth of the programs has been irregular, and measuring the successes or even the magnitude of the aid is most difficult. U.S. aid has been by far the largest in quantity, but not the largest in proportion to national wealth and product output. The effectiveness of this aid in raising living standards within the recipient countries has been lessened because it included large quantities of military goods and services and aid that was apportioned for political purposes. In fact, the incentives for aid programs as far as the donors were concerned have been political and military as well as economic and humanitarian. The unselfish component has probably been rather small. Of the $300 billion or more contributed by the United States, so much can be charged to the costs of war and the rebuilding of warring nations that the aid to underdeveloped countries seems like a pittance (see next section).

The immediate objective here is to examine the status and the prospects of those countries in which economic development and industrialization have been inadequate to provide reasonable living standards for the growing population. The only comparable statistics available for measuring such economic levels are the data for each country as a whole, so the only possible comparisons are by averages for one country as compared with those for another. Such averages fail to indicate the extremes of poverty and riches in each individual country. Nevertheless the contrasts with industrialized countries are startling.

Since the developing nations already contain a large majority of the world's population and are growing much faster than the rest, their highly inadequate living standards represent a great loss to the human race and even a threat to world security. Aid to these nations must somehow promote their development, as the situation cannot possibly be improved by a massive dole. The urgency of a wise aid program can only be grasped as responsible citizens in developed countries become aware of the facts and become convinced that aid must be converted to trade as fast as possible. Many of the earth's great resources reside in the developing countries, and from our experience with the oil-exporting countries we already know how a determined few (in the OPEC) can use this fact for economic blackmail and political threats. If we are to deal wisely with the less developed countries, which comprise the world majority, we need to know more about them (Table 1). In terms of the advantages that we take for granted for ourselves, what is the contrasting status in these countries? Are they very poor? How do their opportunities compare with ours, in education, in health care, in the uses of energy, for example?

We soon learn that the contrasts are appalling. Our average yearly income per capita is of the order of fifty times as great as the individual income of each of the half billion people in the countries listed in Table 23.[39] The developed countries average nine times as many teachers per 10,000 population, much more food (Table 7), and a much longer life for each individual (Table 23). An even greater contrast is the extent to which the brute strength and drudgery of humans and animals has been replaced by other forms of energy. In electric power, for instance, the contrast is over seventy to one. With their 90 kilowatthours per year, they could have a dim light in each home; but none would be left for lighting other buildings, for business or industry of any kind, for outdoor lighting, for radio, television, or telephones, or for any transportation. Perhaps electricity is somewhat of a luxury in a country where there is little industry or commerce, but the same cannot be said of health services. A sample of the contrasts here is given in Table 24.[40] No other kind of service is more important to the individual. Consider the chances of good health care in a poor country where there is only one doctor, one nurse, and ten hospital beds for 5000 people or more. Compare this with one of the developed countries where there are ten doctors, twenty nurses, and at least fifty hospital beds for the same number of people. Here also the doctors have better equipment, and both doctors and nurses have had a better education. This is one reason for the difference in life expectancy. (See also Tables 7 and 27.)

Such advantages as thirty times as many automobiles per capita or eighty times as many telephones seem less important, especially to some who can remember when these were not a part of our good life either. But starvation diets, low life expectancy, crowded dwelling units, inadequate water supplies, and few educational opportunities represent to every American a condition of poverty that no nation in today's world should be required to tolerate. Whatever the defects of the statistics, it is clear that the contrasts between our conditions of life and those in such have-not countries are almost beyond our ability to imagine, especially if we project what these contrasts are going to mean in the future.

Through modern communication the developing countries are becoming aware of the differences. They are demanding more equitable levels and are especially vocal through their plurality representation in the UN. What will it take to ensure that terrorist minorities in some of

[39] Table 23. Average National Income per Capita, 1970: Samples from Each End of the Spectrum.
[40] Table 24. Health Services—Affluence versus Poverty: Doctors, Nurses, and Hospital Beds per 10,000 Population.

these countries do not get the upper hand? We are almost helpless in fighting such terrorism, as has been well demonstrated already.

What, then, should be the international program? Actually, activities to aid developing countries are becoming worldwide and well organized, but a great uncertainty arises as to the eventual results in the face of inflation, higher costs of energy, and of the very high rate of population growth in the underdeveloped nations. Some of them are also spending large sums on armaments and military organization. In addition to food and consumer goods for the large population increases, jobs must be available for oncoming generations.

Perhaps these conditions in the poor nations can best be realized by considering the situation for a single family. For example, an average farm family in the Philippines might include seven children, or a family of nine. When the children are young, this is a heavy load of dependents for the head of the family; and it is most difficult for him or her to save money for farm improvements, machinery, and better methods from a *total income* of less than $150 per month. The family budget for food, housing, necessities, and education is very much larger than it is on a farm where the family size is three to five instead of nine. The number of jobs that must be created each year in a rural-type economy with seven children per family presents a grave problem. The traditional method of subdividing farms and working them more intensively has already gone on for generations; and as the farms become smaller, they become less adaptable to modern, mechanized farming. In areas where there is much rocky terrain, the fields are often surrounded by stone walls and rock piles which represent generations of clearing and building. Here any quick conversion to mechanized farming with its large tractors, combines, and other machinery is quite impossible. In many parts of Asia in particular, farms average less than 5 acres each. Better farming will help out, but industrialization is the only way of providing enough jobs for the overflow of workers from the large farm families generation after generation. Capital for industry should come from savings accumulated by the people, but a good farmer with seven children will certainly need all his savings to maintain and improve farm buildings and equipment. The investment of 10% of this farmer's gross income in industry cannot even be approached. (This figure of 10% is often mentioned as a minimum to help provide jobs for sons and grandsons.) Other possible sources of capital for industry are loans, which must have a solid prospect of being repaid, and subsidies or grants from outside the local community. The suppliers of such grants must have some incentive for providing them. Loans and grants from one country to another are not new, and the machinery for originating and handling them has greatly increased since World War II.

The sample of have-nots in Table 23 represents a population five times that of the United States, but our country today is already using about 30% of many of the natural resources being utilized today on spaceship earth (fuels, for example). If there are already many complaints about scarcities, what will happen if and when these developing countries *do* become industrialized?

The economic accomplishments of a lifetime, or of a generation, or even of a decade, in an industrialized country, are substantial and measurable. Perhaps it takes a decade to introduce the transistor, a generation to develop high-yield, hybrid grain, a lifetime to perfect the motor vehicle. Developed nations are proud of such accomplishments, and confident that others will follow. However, in comparison with undertaking the improvement of the lot of two-thirds of humankind within a generation or two, these are small strides. The industrialized world has been unable to face up to that challenge, especially after realizing that in the meantime the numbers to be helped will at least double, and even now there is not enough food to go around. Some of the aid programs and the development problems are considered in the two following sections.

B.   U.S. FOREIGN AID: POLICIES AND ACCOMPLISHMENTS

The current U.S. aid-for-development programs were a natural outgrowth of the post-World War II relief and mutual security grants and loans of the 1946–1960 period, including the Marshall Plan. In the case of the latter plan, the American people were willing to divert over 2% of the GNP to this purpose. Of this sum 83% was outright grants or gifts and only 17% loans. Until about 1961 the grants continued to exceed the loans, but since that time the pattern has changed, as indicated in Table 25.[41] It will be noted from this table that for the years 1945–1973 the total foreign assistance made available by the U.S. government averaged more than $5 billion annually. This does not include sums from private sources.

In the period 1962–1968, more than one-half of U.S. economic aid was in the form of loans. Much of the 1962–1973 aid passed through the

[41] Table 25. U.S. Assistance and Military Outlays, 1945–1973: Economic and Military Programs.

Agency for International Development (AID). Of the rest, the largest item was for food. The Peace Corps received something less than 2%. It might be noted that, in the earlier postwar years, very large amounts of the total economic assistance, including the Marshall Plan, given by the United States went to countries that had been involved in the war, rather than to the underdeveloped countries.

Yearly data for fully evaluating the aid programs are very difficult to provide. Not only is there an indefinable line between gifts and loans, but many statistics represent commitments rather than deliveries; and irregular inflation has made the monetary measures unsatisfactory. Most statistics are quoted in dollar values as of the year being examined, and conversion to a constant dollar will not be attempted. Percentage ratios are actually more meaningful, for example, aid as a percentage of the GNP for the same year. The aid programs will be considered largely from the U.S. viewpoint, with little attempt to indicate their adequacy from the viewpoints of the recipients.

U.S. programs have been subject to detailed and almost whimsical limitations, specified in the congressional actions. Continuity must be established, however, as there is no question but that the programs of aid to developing nations will be in our own interest for a long time to come, and aid must be increased. There is every indication that today's indecisions are leading the world toward violent upheavals, including mass starvation of some of the fast-growing populations. We are already aware of terrorist activities and the disruptions and fear caused by them, and of the widespread starvation in crowded areas where maldistribution and inadequate food production exist. Even unfavorable weather conditions for 2 years or more, such as those experienced in Africa south of the Sahara, can upset the ability of an entire region to sustain itself.

Gradually the program of aid to developing countries has become international, and to some extent multilateral. The United States has always been the largest contributor and has in fact contributed more economic assistance than all other industrialized countries combined; but we do not rank first in the percentage of the GNP contributed. Of less than a dozen other substantial contributors, the first three are France, West Germany, and the United Kingdom. The United States, like other countries, has preferred bilateral arrangements, that is, each country dealing individually with the recipient countries. Multilateral and cooperative assistance is increasing, however, and the two most important agencies in this field are the World Bank group and the International Development Association (IDA) of the UN.

The less developed countries have almost two-thirds of the world's population, and the assistance provided by the United States, only

5% of the world's people, is small in comparison. It has been very much less than 1% of our GNP; and since roughly one-half of our aid is in the form of loans at interest, and a fair amount of it is "tied money" with the provision that the countries must buy from us, the unselfish help we have been providing is not very large. As to loans, the "soft" loans that are payable in local currency, which is inconvertible into a hard currency, have always been difficult to administer in a businesslike way. When a loan becomes an impossible debt, it may interfere with continuing programs. It is doubtful whether development aid can ever be adequate, but it is a grave question whether the total today is enough to ward off catastrophies, such as widespread starvation and indiscriminate terrorism. It is even entirely possible for one of the less developed nations to build up a military machine that includes nuclear capabilities and then to threaten some of the developed nations. This is one of the dangers growing out of the proliferation of nuclear power plants. The Arab nations are a case in point.

Methods and policies vary greatly among the many agencies involved in U.S. foreign aid programs, public and private. Changes in the names and the objectives of individual agencies also occur, and this makes for confusion in reporting the programs. The three major types of aid are grants, technical services, and loans. Military aid is in a class by itself, because its objectives are political rather than economic or humanitarian; and its contribution to the living standards of the recipient population is very indirect, and often negative. There is much disagreement on how aid to developing countries is to be administered, and not a little pessimism as to the results. High rates of population growth can cancel much of the progress. Futility looms when the great educational tasks are considered. There is a lack of organization due to inexperienced leadership in the recipient countries, especially at the distribution levels.

One of the vexing questions in the case of American aid is: If the United States makes a loan that is used by the borrower for the purchase of U.S. goods, does the United States gain by the transaction? Until the loan is repaid, the transaction is of course incomplete; and the ultimate gain depends on when, how, and whether the loan is repaid. It is often impossible to trace such transactions to the end. If imports and exports finally balance for both countries, the aid is merely an extended business transaction, with perhaps a low rate of interest. If the loan is not repaid, the goods are a gift. In any case, when goods are sent out of this country and paid for with money loaned to the purchaser, the American labor and resources are a temporary gift to the purchaser; and the American workers that produced the goods did not thereby contribute to

American stocks of goods. The workers were temporarily supported by fellow citizens on a sharing basis. Only after the debt is somehow repaid in goods can the shared stocks be replenished.

The final long-term measure of aid for a receiving country is the value of the incoming goods and services as compared with the value of the goods, if any, exported to pay for the imports. An examination of the actual movements of goods representing imports and exports of the underdeveloped countries for the 1966–1971 period seems to indicate that about 10% of the goods imported were not paid for by exports. To the economist this is an unfavorable balance of trade. To the extent that the 10% represents donations from other countries, however, the results are certainly not unfavorable to the recipient country.

What is actually the most effective kind of aid for encouraging economic development in a receiving country? Food and consumer goods are not likely to do much in this direction, nor is the development of the recipient country increased when the goods are handled largely by a temporary staff from the donor. Yet a good organization and some accounting skills are necessary in any large distribution of goods. The development process is a slow one. It calls for education, plus capital goods and the skill to utilize them. In most countries little headway is possible until food production has been improved, so that former agricultural workers become available for employment in industry and in the services. We have learned that some of the most successful activities are those in which we help small communities to provide for themselves better domestic water supplies, streets and drainage, and improved sanitation. Building classrooms and setting up simple labor-intensive industry are other examples of preferred self-help projects. In all cases the objective must be to help people develop on their own. The individual U.S. worker sent to those countries must learn the native language, customs, and standards, and live accordingly.

One field of extreme importance has often been given only secondary consideration, or even omitted entirely in the planning of aid; and this is assistance in improving the health conditions in the recipient country. In many tropical countries especially, the prevalence of preventable diseases is very high. In fact, much of the apathy and inefficiency of the working population can be traced directly to health conditions, not merely malnutrition, but even widespread incidence of debilitating disease. Malaria, hookworm and intestinal worms, diarrhea afflictions, and other treatable diseases affect large percentages of the population. Aid funds spent in raising health standards provide a high economic return; but since the results are hard to measure in numerical terms, as compared with tons of crops or products, direct appropriations for assistance in this field are often meager.

A mix consisting of direct aid together with free-enterprise investment in the receiving country might be of great value, but resistance on both sides makes it hard to get started on any major scale. Many underdeveloped countries have had bitter experiences with foreign developers of industry within their boundaries. Exploitation of native labor, exhaustion of natural resources with little gain to the country, and spoilage of the environment, have created hostility. This has occurred especially in mining and extractive industries, including petroleum.

Crisis conditions for foreign aid programs have developed rapidly ever since the Arab oil embargo and accompanying oil price increases by the OPEC. Deepening economic troubles in the U.S. added greatly to the difficulties. The Russian wheat deal and exhaustion of the long-standing U.S. grain surplus in storage, followed by a rapid rise in food prices in the United States, caused a public reaction bordering on panic. By the time of the UN World Food Conference in 1974, widespread starvation in parts of Africa and Asia was already developing, with threats of more to come.

Poor crops in many places in 1972, in parts of the United States in 1974, and in several other countries in 1975 added real concern. In the meantime, foreign aid programs did not fare well in Congress. Many people insisted that help for the disadvantaged and the unemployed in our own country must take precedence over foreign aid, especially since we no longer have a grain reserve, and since the poorer nations seem to develop an attitude that they should rightly receive certain quantities of free food regularly, even while they are spending heavily on weapons and military establishments in preference to buying food.

When President Ford, in February 1975, announced an increase of two million tons of food to be shipped on a free or a subsidized basis, the aid program was almost doubled without direct congressional action; but Congress shows a negative mood, and is inclined to reduce the aid appropriations.

Emergency moves to meet successive crises fail to contribute to a much-needed long-term plan for aiding the developing nations. Every plan for future development aid for these nations has suddenly become obsolete, because all available financing within most of the countries has been diverted to meeting the increased energy and fertilizer costs for food production. None remains for industrial development. This certainly postpones the day when foreign aid expenditures by the United States can be reduced.

If a developing country has large natural resources, it may grant concessions to foreign corporations to utilize them and thereby spur its own development. Some large Western corporations are wary of becoming involved in the underdeveloped countries, however, especially in projects that call for very large investments. Too many small countries have nationalized such facilities very soon after they started operating.

Domestic troubles have diverted U.S. attention from broader world issues, and many people are poorly educated as to actual world conditions. Humanitarian motives for helping poor countries have been weakened as church and religion have found a lesser place in our lives. However, the theme of interdependence is now getting much support from our leadership. President Ford, Secretary Kissinger, and former UN Ambassador Moynihan not only have emphasized the economic interdependence of the developed and the developing nations, but have in very specific terms declared that the problems of energy, environment, population growth, and food supply transcend national boundaries and are a challenge to the international community. We are the largest and richest member of this community.

C.   DEVELOPMENT PROBLEMS AND THE DEMOGRAPHIC TRANSITION

Will the underdeveloped countries *ever* be able to make the demographic transition from a poor country that is largely agricultural and has a low per capita income, high birthrate, and high death rate, to an industrialized country with a high per capita income, low birthrate, and low death rate?

In summarizing the economic status of various groups in terms of national product per capita, the UN has published data giving the averages for three groups of countries, as of 1970, in then-current U.S. dollars. The figures are stated as "gross domestic product per capita per year in purchasers' values," for market economies:

Developing market economies, $220

Market economies, $1020

Developed market economies, $2960

While the significance of these contrasting amounts cannot be fully appreciated at a glance, they vividly indicate that economic progress from the first group to the third is no small accomplishment.

The industrialized countries are only gradually becoming convinced that major programs of development assistance to the poorer nations are not just a humanitarian gesture, they are actually a necessity for a stable world. Each program of this kind is of course designed to benefit the donor nation as much as possible. Over the past 20 years the patterns of such foreign aid have become well defined, and more-or-less permanent organizational structures are functioning, with the UN as the focus for multilateral efforts. The long-term UN goal of 1% of the GNP of each industrialized nation as a yearly contribution to the economic development aid program has not yet been attained by any large nation, however, and the present level is less than one-half of this. (For the national product of individual countries, see Table 1.)

Because of the diversity and complexity of programs for development aid, it is difficult to present a quantitative summary of all bilateral and multilateral aid, including the trends from year to year. Inflation distorts monetary comparisons, as do the problems of international payments, devaluations of currency, wars, revolutions, dictatorships, and widespread crop failures. The UN tries to provide tabular information, but some nations fail to submit adequate data, and the time lag is always great. Table 26[42] gives pertinent data for 1968–1972, with some separation of statistics for market enonomies, centrally planned economies, and developing countries.

The UN goal of economic growth for developing countries, 5% per year, if ever attained, should enable them to double their total GNP in less than 20 years; but unless their rate of population growth is reduced in the meantime, the product *per capita* will increase very slowly. After the big increases in the price of Arab oil, a 5% yearly growth became an impossibility for the poor nations because their cash and credits for imports were all used up in paying energy and fertilizer bills.

In several of the less developed countries, the GNP per capita is about $100. With a 5% yearly increase in output (in constant dollars) and their present rates of population increase, the prospects are that it will take almost 50 years for such a country to reach a GNP per capita of $400; and this is still way below the poverty level. In the meantime, each highly industrialized nation such as the United States, West Germany, or Sweden, would also increase its GNP per capita, say from $4000 per year to $8000 per year, even at the slower growth rates proposed for the future (as discussed later). Thus the goal of 5% yearly growth in output, which has not yet been attained, would leave the future citizen of these have-not nations with an income only 5% of that enjoyed by the prosperous nations. Will they tolerate this?

---

[42] Table 26. Development Assistance for Developing Countries: OECD and Other Official Programs.

Much development assistance has been in the form of loans, and many of the leaders in the developing countries are fast becoming concerned about increasing debt. In a primitive or marginal agricultural area, where there is not even enough food for an adequate diet, saving to pay interest and repay the principal of loans is a near impossibility. Very often the people live on poor land, and crop failure, due either to drought or to torrential rains which wash away the crops, are not unusual. There are many community projects from which there are no direct earnings to pay even the low interest charges on the loans. These include the building of roads, bridges, water supply systems, and classrooms, and making bricks for public buildings. It might be desirable and necessary to make outright grants for all such nonprofit purposes and to use loans only for assistance in the development of industries. Once a producing industrial plant is established, it could be expected to repay modest loans which helped to finance the start-up costs.

It seems that there is as yet no ready approach by which aid programs to developing nations could be put on a more businesslike basis so that their effectiveness and continuance would be ensured. Nevertheless, the simple facts of population growth in the primitive economies should be sufficient to justify foreign aid programs, especially when these are seen in the context of our own future. The cultivation of friendly nations, direct access to needed raw materials and unique products through trade, prevention of the rise of desperate minorities who may decide that violence is the only way to get attention—these are powerful incentives for the continuation of foreign aid to serve the interests of the developed nations. Besides, there are humanitarian motives that we all endorse.

Some of the necessary programs for a developing nation contribute only indirectly and gradually to its economic gain, hence the ample financing of such programs is difficult. Health care and education are prominent in this category. Although these services are of great importance to every family, a poor nation is unable to establish and support them at an adequate level. Not only are these programs expensive, their success even tends to increase temporarily the national economic burden by furnishing the knowledge and conditions for increased longevity. Better care of infants and the aged directly increases, temporarily, the rate of population growth.

The pitifully low level of health services in some of the developing countries has already been noted (Table 24). Educational deficiencies are not easily measured, but statistics are available on the existing numbers of teachers and students, and some comparisons are given in

Table 27.[43] This table indicates that in the developing countries there is comparatively little educational opportunity above the primary grades—and this is a very serious deficiency when it applies to two-thirds of the human race. Such conditions must be corrected before there can be much progress in economic development. How soon can this be accomplished? Certainly not in a "5-year plan."

There have been many serious studies directed toward the future of aid provided by the industrialized nations to promote economic development in the developing nations. In the United States these studies have not commanded wide public attention and have not resulted in any continuing policy toward a specific foreign aid goal. The reasons for generous economic aid were widely accepted in the days of the Marshall Plan. Both the needs and the results were plain to see. It is doubtful that the average citizen today sees good reason for a large foreign aid budget, especially aid for economic development. The provision of aid funds by the U.S. Congress is greatly affected by pressure from constituents to do more for the less fortunate in our own country.

Among the few congressional leaders who emphasize global interdependence is Senator Hatfield, who has often called attention to our dependence on Third-World natural resources. He has quoted figures for the 1959–1969 period, stating that U.S. investments in Europe of $16.23 billion gave repatriated profits of about 45%, whereas Third-World investments amounting to $5.8 billion gave a return of over 250%. This contrasts with "aid," but such favorable conditions will probably not continue after the poor nations realize how dependent we are on their products.

There are actually good reasons for attitudes of futility with respect to foreign aid, however. As already indicated, any great progress toward adequate economic development in the Third-World nations seems impossible within a reasonable time, even 25 or 50 years. One of the most thorough, yet discouraging, of recent studies was described in the second report to the Club of Rome (1974) (see Chapter 8, Section B). Conclusions were that any substantial reduction in the economic gap between the industrialized and the underdeveloped countries within the foreseeable future is practically impossible unless rates of population growth in the latter are greatly lessened very soon. A reasonable program that would reduce the present per capita income gap, now thirty to one or more, down to five to one or less was studied in various

---

[43] Table 27. World Status of Education, 1970–1971: Teachers and Students per 10,000 Population.

"scenarios." The most promising was a crash program starting almost immediately and increasing in size each year to 2025. The proposed maximum annual rate, year 2025 was $250 billion 1963 dollars, perhaps $25 billion that year for the United States. The total for the 50-year project was about $800 billion.

Such figures are most discouraging. They do demonstrate that to attain any substantial narrowing of the difference in average individual income between rich nations and poor nations will require a much higher scale of foreign aid than anything contemplated today. With the worldwide industrial slowdown the attainment of economic growth for the poorer nations becomes even more formidable.

The UN and its many agencies and meetings have done much to focus attention on the world problem of the less developed countries. The UN statistical reports and yearbooks are invaluable. The many UN world conferences have dispelled misunderstanding through face-to-face sessions and extensive committee discussions. Unfortunately, the haves and the have-nots each tend to segregate, and sometimes instead of constructive efforts toward a useful program of aid to the less developed countries the meetings develop into a political hassle between the two groups. This danger has been further activated by the recent admission to the UN of many small, underdeveloped countries, who tend toward bloc voting with the less developed countries.

A most encouraging promise of U.S. cooperative economic assistance was the 1975 declaration of Secretary Kissinger, speaking to the new nations of the UN General Assembly: "We have heard and have begun to understand your concerns . . . We are prepared to undertake joint efforts to alleviate your economic problems . . . the two sessions this fall could mark the beginning of a new era in which the realities of an interdependent world economy generate a global effort to bring about peaceful and substantial change." Secretary Kissinger went on to say that "ideology is an unreliable guide," that the less developed countries have divergent interests as producers and consumers and as exporters and importers, and that these "cannot be resolved from the unreality of bloc positions."

A final conclusion must be that wise foreign aid planning is definitely of comparable importance to world planning in other fields, such as energy supply.

D. WORLD ECONOMIC DEVELOPMENT—CONTRASTS

The following three chapters deal largely with future prospects in the United States for the next 10 to 25 years. These prospects are related to our experiences of yesterday and today. Whatever attention has been paid to other countries throughout this book is expressed largely in terms of contrasts with conditions in our own country or in relation to the part the United States plays in the world scene. This narrow concern must suffice, because of the specific and limited objectives of the book. Admittedly, this gives an inadequate and one-sided view of the rest of the world, especially since we constitute little more than one-twentieth of the present world population. Nevertheless, these contrasts are also important as reminders that our problems are modest by comparison.

There are many reasons, both selfish and unselfish, why we should strive to give effective assistance to peoples whose economic and personal struggles are in such pathetic contrast with our own. These needs and contrasts have been given some attention in this chapter (and in others), but a short summarizing discussion seems in order.

Our planet's natural resources are widely dispersed, and there are few remaining good-resource areas where the population is so sparse that people cannot utilize nature's gifts. Most natural resources are underutilized, however, as a result of human shortcomings—lack of knowledge, initiative, and organization. Our procreative urge plus our gregarious nature have gradually resulted in centers of overpopulation. This is something relatively new. In terms of human history the United States is a new country, yet between 1776 and 1976, *world* population grew to almost five times the size it had attained in all prior human generations. Our own population grew from less than 4 million to about 214 million, about sixtyfold. No wonder we face certain problems with which history gives us little help.

Quantitative measures of national economic development are elusive, and not convincing. Nevertheless, one must agree that such realities as the per capita food supply, health care, education, and total income, do index the level of well being in a country (see Tables 1, 7, 20, 23, 24, and 27). The contrasting national or regional averages unfortunately do not show the great differences between rich and poor within every country, nor do they remind us that in very poor countries some of the population is barely above the starvation level.

One striking difference between the rich and the poor countries is in the number of people employed in agriculture. In an industrialized and economically developed country only 5 to 10% of the working population are directly engaged in agriculture, while in a developing country the percentage is well over one-half, probably 80% or more. In the latter country there may even be high unemployment, because there are no opportunities for workers to shift to industry. A developed country has a very high percentage of gainfully employed, including women, a high production output per worker due to advanced technology and the use of mechanical and electric power, and a large middle class with capabilities in specialties of all kinds, including health care. These conditions make possible a high rate of savings and investment from annual income, both individual and business. Thus the industrial production plant continues to improve in quality and increase in size.

The gap in living standards between rich and poor nations can be reduced very little, even in a generation or more, by any means available within the poor countries themselves. Not only is the present per capita imcome in a developing country only about $100 per year as compared with perhaps $3000 per year in an industrialized country (1970 dollars), but the more rapid population increase in the poorer country requires a doubling of supplies of food and consumer goods within a single generation.

Suppose that in both a rich and a poor country the rate of personal saving is 10% per year, and that $2000 is required as minimum investment, in tools, equipment, or stock of goods, to equip the simplest type of nonagricultural work place. The average worker in a developed country could pay off this capital investment in about 7 years, while the worker in a poor country would be accumulating only $70.

Thus most of the capital for economic development of all poor countries that do not have large exports must be provided in some way by the industrialized nations. Capital accumulation is a problem for the entire underdeveloped world, and the UN announced in 1970 its "second development decade," 1971–1980, and planned an "international development strategy." In 1974 a review committee reported that this strategy "has not yet taken hold with anything like the pace needed," and Secretary Waldheim made two appeals for help on an emergency program. Response was small, and some countries, including the United States, decided not to participate.

The underprivileged half of the human race now has a forum in the UN General Assembly in which it can present its case, and each of the nations has a vote that counts the same as the vote of the United States. Comparatively, their economic situation is becoming worse year by year, rather than better. There is every need for the powerful nations to curb their own impatience with the several unwise resolutions and actions of the new UN majority and to use all possible wisdom in promoting economic development.

The developing countries represent a great reservoir of manpower, and they want to help themselves if they only can learn how. Their lands contain vast stores of essential raw materials in need of development and marketing. But the difficulties in promoting the economic development of these countries seem to be increasing, especially with worldwide inflation and quadrupled oil prices.

The UN General Assembly has called for special measures on behalf of the least developed countries, or sometimes, "fourth world." In these countries the contrasts with our standards are most extreme, and the natural resources are not impressive. Their subsistence farmers seem most resistant to change, and the economic conditions are the most hopeless. In spite of their plight these countries were receiving much less aid per capita than those in other underdeveloped countries. Almost a billion individuals in these poorest countries have been living in dire poverty, partly because of their resistance to anything new, and also because of the inexperience of what central government they have. Even so, such central governments are aware of UN economic statistics which serve as a continual reminder of their unfavorable status compared with the industrialized nations. As a result of their knowledge of these contrasts there is increasing agitation on the part of the have-not nations for a more equitable sharing of the world's goods. It is not only the oil-rich nations that want a higher return for the valuable raw materials that leave their shores. Other resource-rich but economically poor nations are learning to increase their demands too.

Sweden in 1974 accomplished the feat of surpassing the United States, attaining first place in per capita income among industrialized nations. In addition, she was the first country to attain the OECD goal of 1% of the GNP for development assistance. Her foreign aid appropriations represented 3% of the Swedish national budget.

The 1974 statistics of per capita income (*World Bank Atlas,* 1975) listed the six highest among the industrial countries in this order: Sweden, Switzerland, the United States (all in the $6600 to $6700 range), Canada, West Germany, and Denmark, This order will probably change again when business picks up. These Western leaders were well behind several of the OPEC nations, who were above $10,000 per capita.

The U.S. Congress has been greatly displeased by certain actions of the UN General Assembly since its majority of small nations began to assert themselves, and is inclined to keep the U.S. share of assistance much closer to one-tenth of 1% of the GNP. It appears that Holland, Norway, France, and Denmark will soon meet the modified goal of 0.7% of the GNP. Our contribution approximates one-half of 1% of the federal budget as compared with Sweden's 3%.

U.S. actions affecting world economic development are nevertheless the most weighty on the entire scene. It is important that the U.S. citizen have knowledge of the contrasts between our relative contributions and those of some other countries, or there will be little chance of upward revision of our share in the development efforts. The problems of the underdeveloped world are staggering. They are increasing, and they will not "go away" while we insist on minimum participation. In this "one world" we are all interdependent, in terms of food and energy, but also in terms of population growth and conservation of environment and natural resources.

More attention to the entire subject should convince us that our greater participation is a challenging investment in the future development of vast populations, resources, and trade, and not merely a dole to provide relief. The younger generation is beginning to appreciate this.

Chapter Six

Past Expansion Versus Slow Future Growth

A. THE TWENTIETH CENTURY; THE U.S. PAST AND FUTURE

The twentieth century, with its wars and depressions, has not by any means furnished 75 years of steady economic growth at recent rates. Familiar annual appraisals of the economic scene such as those in corporate reports or in the January issues of the business press are naturally overinfluenced by the immediate past. A safer estimate of what the future has in store is obtained by occasional long-term examinations of economic experiences with accompanying historical data such as the statistics shown in Tables 5 and 28.[44]

Past experience, as presented for the United States in general terms in these tables, will supplement the impressions already obtained from previous chapters which dealt with specific critical fields. While the facts regarding food and energy supplies, our relations with the "developing two-thirds," and the capacity limits of spaceship earth are presented as critical factors in our economic future, it should be recognized that there are also persistent long-term trends of a general nature. Moreover, there are always possibilities of completely upsetting experiences, such as a nuclear war, or even a combination like the long depression and dust bowl experiences of the 1930s.

General observations and data about this century with which we have had some personal experience are an appropriate preface to any speculative predictions and extrapolations for the next 25 years. The

---

[44] Table 28. U.S. Growth by Decades, 1900–1973: Ten Economic Indicators.

following sections are concerned with some of the visible trends of today, their future risks and limits, and the changes in life-style and philosophy that will probably be required. Such a shift from facts to anticipation gives the reader ample room to frame his or her own argument.

Only a sketchy sampling of U.S. growth and economic conditions is provided in Tables 5, 10, and 28; but they show that rates of growth have been startlingly high, especially since 1950.

Overall economic growth as indicated by the GNP increased fivefold in the first 30 years of this century, while in the 30 years from 1940 to 1970 the increase was almost twice as great (Table 5). Personal income per capita, in constant dollars, grew over 2.5 times as fast during 1970–1973 as it did during the 1950s (Table 10). In constant dollars the income of each person in the population averaged five times as great in 1973 as it was in 1900 (Table 5). Our faith in a good education ascribes some of that economic gain to the fact that almost 4% of the population was in college in 1973, twelve times as many as in 1900.

One startling change has been the shifting of much economic activity to the federal government, with a consequent increase in taxes. Federal expenditures increased almost 500 times during 1900–1973, while the population did not even triple, and federal debt per capita increased over a 125 times. The economic choices for the individual, in spending his or her income, have therefore been greatly lessened. Each individual is now paying interest on well over $2000, their share in the federal debt (Table 28), and a substantial fraction of every family's income goes to taxes.

A change that is not often mentioned for twentieth century United States has been the decrease in birthrate. It was 30 per 1000 during 1900–1920, but only 15 per 1000 in 1973.

The several interim emergencies and triumphs are admittedly not shown by a decade sampling of U.S. growth and economic conditions for 1900–1970; but attention is at least called to several persistent long-term trends such as inflation, debt, employment conditions, and standards of living. They also emphasize the improbability of accomplishing such objectives as zero growth or zero pollution. Only with cooperation from the entire population can changes in major trends be accomplished, hence the importance of knowing the facts of past economic performance such as those presented in the many tables herewith.

After the rapid expansion experienced in recent decades, any shift to slow growth will be difficult. Progress can be expected to be painfully uneven, even if war and terrorism can be avoided. Depression symptoms have already been experienced. Unfortunately, these relatively sudden slowdowns are considered highly newsworthy by the press and television, and panic is thereby incited. This lessens stability, especially that of the stock market. Economic values and assets are thereby greatly distorted, and temporary fluctuations magnified.

The projections here selected as the probable and recommended rates of growth to the year 2000 have already been presented in Table 19, and they should now be reexamined and compared with the past performance shown in Tables 5, 10, and 28 in an effort to determine what expectations are realistic. Lower growth rates in the future must now be anticipated in almost all economic activities.

Capital and manpower needs will impose real limitations on many proposed economic plans such as Project Independence, full control of pollution, and expanded foreign aid (see Chapter 7). In fact, such concurrent efforts added to those intended for social progress at home, and our vast military expenditures, each call for large capital outlays and skilled manpower services. This competition with its various demands for economic growth runs counter to the slow growth requirement and also conflicts with the demands by present and future consumers for improvement in their standard of living. The conflict between the demand to accomplish many specific nonconsumer objectives and the limitations that restrain such accomplishment are of course nothing new. This is the very nature of the struggle for economic progress.

Much has been said in previous chapters about why the pattern of rapid U.S. growth must be changed, but it is not easy, in view of our economic accomplishments during 1950–1973, to abandon the long-term conviction that progress depends on rapid growth. Have we no choice now but to approach zero growth as soon as possible?

If this question were to be answered by a general poll, the result would probably be a wholesale denunciation of the new aim of slow growth. While rapid population growth is easily accepted as the main reason for the ills of the underdeveloped countries, most U.S. citizens continue to regard fast growth as the necessary factor for economic recovery and good times. True, this viewpoint is being modified somewhat by environmentalists who are already having a marked influence on future planning in the United States.

Enthusiasts in general, and environmentalists in particular, are inclined to advocate extreme measures or drastic changes from what has been accepted practice in the past. While this attracts much public attention, it is significant that such a comprehensive study as that reported by the Commission on Population Growth and the American Future (footnote 57) emphasized that there is actually a wide choice of when, where, and how the lowering of both population growth and economic growth can be adopted (see Chapter 2, Section G). They said in conclusion, "The central issue emerging from this study is one of choice."

In this report, a main conclusion under the heading "Population Resources and the Environment" (vol. 3) was that it would be quite possible to find ways to cope with continuing population and economic growth in the United States for the next 30 or even 50 years. But after reporting at some length on our past experience and the prospects resulting from each of the likely growth scenarios, the commission concluded that the social costs of rapid growth would be very high. Many of the options that are attractive today would be gradually closed off. Rapid growth would cause an irreversible change toward much less personal freedom, call for extensive public regulation, aggravate social problems, and direct much capital to new and more expensive methods for increasing supplies of energy, metals, crops, transport systems, and services. It was concluded that controls would be required to channel major efforts into mere expansion rather than improvement. There would be more filling out of forms, arguing with the computer, appealing decisions, and correcting environmental damage. A final result would be the weakening of the position of the United States with respect to other countries, that is, there would be less independence to make our own decisions in the world.

American economic expansion in the past has been phenomenal, but the changes that will be required to adjust to the demands of the future seem to be many and drastic. The limits on future growth are definite. The roles of the participants, government, business, and the public, remain to be worked out, but some guidelines are beginning to be evident.

B. NEW OBJECTIVES FOR PRIVATE ENTERPRISE

In America, our business-oriented civilization is inclined to panic at any experience of slowing growth. Success has always been conveniently measured in terms of growth, whether the private enterprise economy is considered a "competitive" or a "profit" system. Endless financial pages, weekly magazines, and business and government reports faithfully list, analyze, and comment on the many indicators that daily monitor the quantity status of business: sales volumes, car and truck loadings, fuel and energy consumption, housing and construction starts, and many others. These must all be increasing to keep the nation feeling good.

One source of the increase, high birthrate, is taken for granted, but seldom mentioned. (Now the birthrate is declining.)

Even the service professions such as education and health care have been geared to quantity production—more students, more and larger schools, larger classes, more patients per doctor and per hospital, businesslike operation, and production-line methods with emphasis on quantity. Financing is semiautomatic—more taxes, more payroll deductions, more insurance. Some changes are going to be necessary. The greatest difficulty will be the adjustments in attitude and objectives, the shift away from domination by a single-minded striving for more and more *quantity*.

A new philosophy and approach is needed, a change in emphasis toward accomplishments for which a growth in quantity or numbers is not necessary. What about emphasis on *quality* instead of quantity? What about better service and more attention to the quality of life? There are no easy measures of these, and advertising themes would have to be changed.

We, the privileged citizens in the industrialized nations, comprising less than one-fourth of the human race, are accustomed to consuming 50 to 90% of each of the world's most valuable nonrenewable materials, especially fuels and metals. Much of our supply we obtain at bargain prices, largely through our own initiative in the locating of sources in distant countries. Unfortunately (for those countries) the bargain prices we pay are possible only because their populations subsist on a living standard less than 10% of our own, and of course the producers are paid accordingly.

In the United States the pattern is even more extreme than in the other industrialized nations. The GNP for the rest of the *world* is only about twice that of the United States. With only about one-twentieth of the world's population, the United States consumes about one-third of the world's valuable energy and nonrenewable materials. This represents not two to three times our even share, but in many cases almost ten times our even share.

Why should we consider a change in our fortunate industrial position? In addition to the troublesome problems of overcrowding and of pollution in our own country, plus the developing scarcities, there is another very good reason. Modern communication, for which we are largely responsible, has dispelled the ignorance of the have-nots. Many of them are becoming fully aware of these disparities. In the meantime, among the items we gladly sell to them are the most sophisticated of modern weapons. Thus is created a situation of potential terrorism by radical minorities who plan to show their attitudes toward us. As we have already learned, we can by no means cope with planned terrorism and sabotage. Thus unstable situation demands attention—and soon.

A decrease in population growth in the United States is necessary in order to control some of the problems ahead of us, but ZPG is unattainable very soon; and perhaps this is fortunate for industrial and financial planners, managers, and advertisers. The U.S. population has recently been growing faster than that of most industrial nations, the increase per decade from 1900 to 1960 having ranged from about 15 to 21%, except during the depression decade of the 1930s. During the 1960s the increase fell to 13.4% for the 10 years, and the U.S. Census Bureau estimates that it will not exceed this rate for the rest of the century. Even if birth control and other means are unexpectedly successful (1.8 children per prospective mother), the growth rate per decade will average well above one-half of the increase during the 1960s, and that is nowhere near ZPG. This is because of the large percentage of women of childbearing age.

An early adjustment to one-half the accustomed yearly increase in numbers of customers will call for some major changes in business and governmental planning. When combined with increasing scarcities and mounting demands from the poorer two-thirds of the human race, America faces some large modifications in life-style very soon; but this is a real challenge, not a calamity, and this is where changes in objectives and philosophy become necessary. What will be the nature of the changes?

C.   INFLATION AND RECESSION

Real success in coping with continuing difficulties in energy supply, control of the environment, food for growing populations, and world development depends heavily on a normal economic climate. Large distortions such as excessive inflation, a sharp business recession, and massive unemployment shift all attention to the immediate emergencies. Efforts toward long-term gains are left almost unsupported by business, government, and the general public. It is fitting in this book, therefore, to include the following short section discussing ideas about the causes and remedies for the economic emergencies that started in 1973, specifically, inflation and recession.

During the recessionary emergency there was much surprise that inflation continued in a period of depression and growing unemployment. However, the inflationary habits acquired during a period of rapid growth are not easily changed, or even recognized. Growth and inflation are almost accepted by business as the normal condition.

Preceding any discussion of inflation in the United States, we should examine the data on actual price fluctuations. Table 29[45] tells the inflation story since 1900, presenting it in terms of 1970 dollars. In this book, where comparisons are made in constant dollars, the 1970 dollar is used, usually based on the wholesale price index (Table A). In the discussion that follows, the contrast between zero inflation during 1960–1965 and a yearly loss of 4 to 10% in the value of a dollar from 1968 to 1976 is of greatest significance.

Once the business community develops the expectation of continuing price increases, its actions in that direction are almost automatic, and hard to stop. Marketing managers at the retail level carry out more-or-less regular price markups, and labor leaders try to keep their demands for wage escalation just a little ahead of the expected increases in the cost of living. Even when sales fall off, prices may level but they are not reduced. Unemployment appears, but labor continues its emphasis on increasing rates of wages and benefits, even encouraging overtime work. Unemployment and jobs for incoming new workers receive minimum attention. The public may call on the government for price and wage controls, public job programs, and expansion of the money supply.

If a downturn occurs, business soon curtails its capital spending, because expansion is not needed and prices look too high for investment or even replacements. Low-level jobs are eliminated first, and the minimum wage is considered too high for inexperienced workers. New jobs become scarce. New workers looking for skilled jobs may try to wait out the recession, because on-the-job training is no longer available and schools seem too expensive. The unemployed in the skilled categories shun menial jobs. They much prefer to keep their seniority and collect unemployment compensation, especially as long as it pays 90% of their usual rate.

These actions are prompted by attitudes and habits acquired during growth and prosperity, and they would be very different if they were based on a careful analysis of the causes of inflation and remedies for recession. Some of these causes and remedies are not difficult to understand and, if they were better known to managers in business and servants in government, the control of inflation could be much more

[45] Table 29. Inflation: The U.S. Dollar, 1900–1975: Equivalent Values Expressed in 1970 Dollars.

successful. The old reliable laws of supply and demand are still very useful when dealing with the causes of inflation or prescribing steps to alleviate it. Any departures from free competition are inflationary.

Continuing increases in the costs of necessities are most serious when a recession begins. One difficulty is that during prosperity so many new things move into the necessities class. This applies to everyone, whether or not their occupation or product is directly related to goods or services in the necessities group.

Consider for a moment the total demands for food, apparel, cars, home appliances, education, entertainment, health services, and vacation trips. These are considered necessities by much of the U.S. population, but at any given time their supply is rather fixed. With the growth of population and prosperity in the United States during 1950–1973, there were more and more opportunities for young people to select a vocation that had nothing to do with producing food, clothing, shelter, or any other recognized necessity. Many families, even those employed in non-necessity vocations, rather than forego anything in the necessities class, will pay a higher price for it. Thus the demand for necessities keeps growing, the supply is rather tight, and prices keep increasing, but without much resistance from users. This is real inflation.

There has been a surprising expansion in nonnecessity vocations. The largest numbers are in *security* forces. This category includes much more than the police, firemen, and members of the army, navy, and air force serving in the United States. After the great diversions of personnel for the Korean and Vietnam wars, we continued to station large forces in Europe, Korea, and elsewhere. Millions are engaged in military operation and administration, in military research and development, and in the manufacture of military equipment, not only for our own forces and bases, but for foreign military aid and for large foreign military sales. Other rapidly expanding nonnecessity vocations are involved in environmental control. Luxury goods and services have recruited many, and the spurt in recreational activities since 1960, for example, boating, winter sports, and travel, has required operators, service employees, and makers of equipment.

Government employment grew more than 100% from 1950 to 1970, while population growth was only a third as great. Social welfare increased tenfold during 1950–1974. In the food and clothing industries, two basic necessities, employment increased at almost the same rate as population growth, while the number of people involved in most of the nonnecessity occupations just mentioned has at least doubled. In addition, the number of people employed in providing 1974 U.S. exports was almost four times the number similarly employed in 1960. Many foreign facilities of U.S. multinational corporations were built and equipped largely by U.S. personnel. Since the definition of necessities is those goods and services that *everybody* needs (and insists on), these are typical overdemand and undersupply situations with heavy upward pressure on prices.

A related cause of inflation has to do with the method of payment. On a pay-as-you-go basis the total price inflation would be modest, as it was in any single year from 1950 to 1968 (averaging less than 2% per year). But in the decade 1965–1974 total debt per capita more than doubled. People are inclined to blame government debt, but private debt was three times as great as public debt and increased much faster. Money and credit were thus greatly expanded, and this is the classical definition of inflation: too much money chasing too few goods (see the dictionary).

Inflation increases replacement costs for equipment in industry and, if replacements are postponed, plants deteriorate, reducing productivity and output, thus causing more inflation. Unemployment also increases inflation, because the unemployed are still consumers; they spend money, but they do not produce goods, hence they must be supported by others, by borrowing or through government programs, again probably financed by borrowing.

Some financial analysts explain the causes of inflation entirely in terms of the money supply. Too much "printing-press money" is an old theme. Today of course, most money is in the form of credit, and there is little printing-press money. This credit includes bank loans and installment accounts and all government borrowing as well.

If there is too much money, the Federal Reserve can *decrease* the money supply. This affects the decisions of both buyers and sellers, and also business volume and interest rates. If recession develops and unemployment increases, there is pressure to *increase* the money supply, and to reduce taxes, but at the same time to increase government spending on relief and public jobs. This means government borrowing, which is inflationary. When inflation and recession occur together, the solution by manipulating the money supply becomes a conflict. It is necessary to turn to basic causes in terms of goods, services, employment, and business activity, plus the intangible elements of public confidence and business sophistication.

If a buyer lacks confidence in the future and his or her resources are limited, he will refrain from buying or investing. However, members of that substantial group whose resources are adequate in the form of savings and money in the bank will tend to increase their buying during inflation, for two good reasons. One impulse is to buy everything needed

before the price goes up, and the other is to conserve long-term resources by buying and keeping things of high intrinsic value, land,
artistic goods, precious metals, diamonds, gems.

The business sector is very sensitive to shifts in the attitudes of
buyers, sellers, and bankers. The profit system responds quickly to
incentives and rewards, and to temporary shortages and excesses. The
resulting actions make for short-term stability and protect the year's
profits, but they do little to correct long-term imbalances such as inflation, unemployment, and excessive spending for uneconomic purposes.
The public sector is also responsive to the influences and emergencies
of the day. Temporary emergencies are multiplied by inflation and recession, with the result that too many government executive actions and
legal remedies apply merely to a one-time or immediate correction. Thus
the reactions of the individual citizen, the business community, and the
government all center on short-range objectives, while long-term improvements are neglected.

Both inflation and recession cause serious social misfortunes in spite
of all efforts to the contrary. One of these is that they bear the hardest on
those who can least afford it and cannot fight back. Personal hardship
comes mainly to marginal and part-time workers, unskilled workers,
young and new workers just hired or not yet working, older workers, and
all those who are retired. Many are affected by cutbacks in government
assistance programs. Society loses the possible contributions of any
and all that are unemployed, but they must still be supplied with necessities. Individuals lose their courage. A few may resort to violence or join
antisocial causes.

It is true that Congress has made some efforts toward programs that
will alleviate such hardships. Support is lacking, because while unemployment and inflation are acute misfortunes for some, they are almost
nonexistent for many others.

Individual concerns about inflation tend to obscure its broad threats
to our entire social organization. Among those who profit from inflation,
there are a few who consider it a benevolent force and advocate an
inflation economy. It is easy to understand, however, that when inflation
continues at a yearly rate that approaches or exceeds the rate of return
on savings, the results spell disaster. It then not only discourages the
accumulation of savings, but also undermines the financing of educational, charitable, and nonprofit activities by endowment income. It
threatens pension systems. Such inflation deprives the retired citizen
and the widow and her dependent children of all income from savings
and from bequests made by family providers.

These social misfortunes cannot be tolerated, and it becomes of the utmost importance to attain widespread agreement on the causes of inflation and general cooperation in its abatement. Since inflation is a world problem of long standing, and since nature makes occasional contributions to it, such as crop failures over a wide area, complete success in its elimination is not to be expected. It is in order at this point, however, to make a summarizing list of goals and corrective actions.

Since recent U.S. experience has involved continuing inflation accompanied by business recession, and the causes and effects of the two are interrelated, the following list of desirable measures and goals intentionally involves both.

1. Free competition is essential. Every effort should be made to ensure its full operation.

2. Great emphasis should be placed on increasing productivity per work-hour, both by encouraging capital improvements in productive plants and by resisting wage increases, fringe benefits, and work rules that lower product output per unit of cost.

3. A widespread educational campaign should emphasize conservation versus waste, especially as regards energy and materials. The superior performance of several other industrialized countries with respect to more economical use of resources should be emulated by the United States, as should their higher rates of personal saving and investment for the future.

4 Productive investment for modernization in industry should be encouraged by all possible means, including tax incentives.

5. Government budgets and expenditures should be required to conform to a pay-as-you-go plan, meaning that there will be no escalation of programs without prospective savings or tax incomes to support them. This includes defense spending.

6. Inducements for personal expenditures out of *future* income should be restrained. This may involve the return of some emphasis on thrift. Educational and advertising persuasion encouraging installment buying would need to be reversed. Advantages to both seller and customer of purchases for cash need to be reinstated. The pay-as-you-go decision is defensible entirely on economic grounds.

7. Public jobs should be a last resort and be so organized that they are less desirable than private employment. Preference should be given to incoming youthful workers.

8. All regulations by government fiat that tend to increase inflation should be frequently reviewed, with provision for modification or suspension of the regulation when necessary. This included minimum wage

and other employment rules, transportation fare regulations, and environmental regulations.

9.   Coercion by strikes and violence should not be allowed to result in inflationary concessions, especially in the public sector. Overtime pay should be eliminated during high unemployment (say, over 5%).

10.   Welfare benefits must somehow be confined to those who need them.

11.   During high inflation special attention should be paid to securing a zero balance of payments in foreign trade. A large export surplus is not a favorable balance when it is a (temporary) gift of goods to receiving nations at a time when our own shortages are causing domestic price increases.

12.   Reliable and detailed statistics on price fluctuations should be collected and supplied to business and to consumers at frequent (weekly) intervals. This will help to prevent unjustified price increases.

13.   All government borrowing in anticipation of future income should be subject to extreme limitations. (Cite New York City as "exhibit one.")

14.   Escalation of either prices or wages in anticipation of *future* inflation should be restrained.

15.   Unemployment benefits should be modest, both in amount and duration, so that they do not interfere with the incentive to work.

The above list is illustrative rather than adequate or complete. Some of the proposed actions will be challenged. On others, there may be room for argument or disagreement, especially in view of the methods required to accomplish them. Implementation of some of the actions listed calls for more sophistication than is usually demonstrated by business or government, but disagreements should at least result in compromise measures that ease some of the ill effects of inflation and recession.

Highly active efforts toward solution of the problems of energy supply, environmental protection, food, foreign aid, and slow growth can be expected only after the acuteness of inflation and recession have been fully dulled, hence the inclusion of this section in the present book.

D.   LIMITING THE ROLE OF GOVERNMENTS

Many loyal U.S. citizens are very apprehensive of any role whatever for government agencies in controlling the output of goods and services from the private business system. It has been recognized only recently that the guarding of air purity by pollution control is hardly any different from guarding the supplies of pure water, or disposing of domestic waste; and these have long been assigned as local government functions. Pure food and drug laws have widened into general consumer protection. The general field of assisting and protecting the individual has now become much more important and in many cases merges with that of assistance to business and industry. In such areas as rural development, housing, transportation, and job training, the benefits to the individual and those to business are almost of equal importance.

For any discussion on government roles, reliable information on past and present conditions must be the starting point. Table 30 gives the percentage distribution of federal government outlays over the period 1960–1974.[46] It will be noted that the fastest growing areas of U.S. government activity are those that contribute most directly to the welfare of the individual. A similar picture, obtained from an outgo or expenditure analysis of the GNP, is presented in Table 33. Here the percentages are those of the total GNP rather than of total government expenditure. This approach again emphasizes the importance of the individual. It also shows that state and local government expenditures, which serve the individual more directly than federal expenditures, have increased even more rapidly than those of the federal government.

The supplies and prices of energy and food are no longer adequately regulated by the law of supply and demand. As long as very limited exports and imports were involved, private development and marketing were effective and had many advantages. However, private exploitation of vast natural resources, private companies furnishing essential utility services, and private control of food exports and energy imports have often resulted in wasteful competition, depletion of domestic resources, and over- or underpricing of products or services. Gradually government regulatory commissions became a necessary adjunct to private operations. One regulatory body after another was set up and charged with protecting the public interest, for instance, the FPC, the Interstate Commerce Commission, and the Federal Communications Commission, in addition to a great many state and city agencies.

Energy and food problems are now quite beyond the control of the private sector, especially in view of the fact that they are not simply domestic problems but are of vital importance today in the development of almost all countries. When our domestic supplies of petroleum, natural gas, and certain metals and minerals ran short, we imported

[46] Table 30. U.S. Government Outlays for Designated Programs, 1960–1974: Percentage Distribution of Federal Government Expenditures.

them on a larger and larger scale. We even obtained the imports cheaply, sometimes by exploitation in a nonindustrial country. For the supplying countries, the opportunities to export were a source of great advantage in many cases, affording local employment and at the same time importing valuable technology for future use. The supply, demand, and pricing of food crops have for generations been beyond the control of producers, and frequently have worked to the disadvantage of consumers. The extensive U.S. farm programs have been reasonably successful for encouraging production and keeping the farmer's income on a parity with others. But now a new governing factor emerges, a continuing and worsening world shortage of food.

Control of U.S. energy supplies is only the first of several problems of resource control to become acute. Others will include food supplies and prices, both domestic and export, environmental deterioration, and supplies of petrochemicals and many metals and minerals. The fields of difficulty are all closely related. For example, any control on the supplies of petroleum and natural gas for fuel will affect air pollution regulations, the fertilizer industry, hence the farmer, and the raw materials supplies for plastics, synthetic textiles, and many chemicals and drugs.

While controls by market forces are the most effective and self-regulating, and government efforts at control have been inept in many respects, neither consumer choice nor the profit motive is adequate to meet the new situations that deal particularly with the limited supply of nature's bounties and the rapid increases in human and animal populations. It becomes very necessary to improve our public regulatory system and to devise incentives for meeting long-range needs. The ordinary American citizen has not been aware of how demands have overrun many U.S. resources, or of the extent to which the rest of the world now depends on us, especially for food.

The fact that energy and food supplies have become critical worldwide is not a dire threat to the sacred rights of property and its private management, but rather a welcome challenge to resourceful Americans to modify our system to meet the new demands.

No sudden switch to new policies will solve our difficulties. The transition to more appropriate control will be gradual, with much trial and error and considerable suffering as a result of errors. The complexities of energy supply-and-use regulations are a good example. Any form of Project Independence to expand our domestic energy supplies cannot be widely effective before 1985; so interim policies must be adopted, including reductions in demand and use for almost all fuels. Personal consumption of fuel and electric power accounts for about one-third of our energy bill; and another one-third is used in the production of goods and services for consumers, including public transportation. A first step in the reduction of fuel use has naturally been the reduction of this two-thirds that goes directly or indirectly to the consumer. But a high-income family uses much more fuel indirectly, through goods and services, than it consumes through direct use as gasoline, heating oil, and so on. Direct fuel taxes therefore penalize the low-income family. Yet it is almost impossible to devise a tax or regulation that will materially reduce the large indirect fuel use by the more wealthy half of the population.

During the early period, say the next 10 years, while consumers are being taxed to control national fuel consumption, great amounts of capital will be required for developing new energy sources and also for additional farm capacity (see Chapter 7, Section A). Foreign aid funds must be increased to ward off starvation in poor-crop areas. Here again private enterprise incentives promise to be inadequate unless supplemented and directed by public measures. In turn, these cannot succeed without wide citizen support, which must depend on a vast program of public information and education. Neither Congress nor the taxpayer will cooperate in regulatory measures by taxation, quotas, prohibitions, penalties, price controls, rationing, or other means unless they are convinced of the need and desirability of such measures.

Inspite of the need for regulatory action, it seems unlikely that government control and regulation will become excessive, given the mood of the powerful American business community. There is probably more danger that business people will not be adequately informed as to the needs and objectives of the regulation and will be able to block the measures or see to it that many are not initiated at all. Here again is a great educational challenge to those who grasp the true situation and have examined the many problems involved.

In a depressed economy there is always much agitation for the government to "do something," and the advocates of government planning have of late been highly vocal. They are held in check by those who are convinced that government interference is bad and that moderate unemployment and hard times are the best cure for a depression or other maladjustment. A struggle between the two factions results in inaction, or at least delay; and in the meantime, small pressure groups that are being hurt may become destructive. An example is that of farm groups that emphasize their complaints by refusing to plant crops and by destroying animals or pouring milk down the drain, in spite of widespread starvation in distant countries.

A compromise solution is necessary. It can only come about through education that lays to rest the arbitrary dictum, "no government interference anywhere," and emphasizes the world-citizen viewpoint. It must be recognized by all that the profit motive is inadequate when it comes to feeding starving people or reducing the great waste of oil and gas resulting from habits developed when fuels were cheap. Yet the farmer and the oil field operator cannot operate with continuous losses. Government support, guarantees, or outright assistance must be resorted to when necessary, temporarily diverting money and effort into the areas in which necessary public policy results in private hardship and loss. The main precaution must be to keep the arrangements temporary. There is little doubt but that more government regulation is in prospect. Our system of checks and balances seems adequate to contain it, but not without painful delays.

The international situation is something else. There is as yet no permanent and successful international agency, although some operations under UN auspices are very valuable, the FAO for example, and the various conferences on population, food, the oceans, and so on. Ad hoc committees and commissions become necessary. Working in cooperation with existing agencies such as the World Bank, there is every expectation that they will become increasingly successful and that the threatening long-term crises of population growth and food and energy shortage can be peacefully alleviated.

F.   FUTURE BILLIONS FOR NEW U.S. PROJECTS

As already indicated, there are several factors that will limit future growth, but mainly the earth's finite resources and environment. A slow, organic growth is here advocated, to forestall future crises, but many people insist that slow future growth would itself be a calamity. Even President Ford, in presenting his budget for fiscal 1976, suggested prospects of continuing steady growth of the GNP well beyond that of the past, namely, 6.5% per year after 1977. This might prove to be the ultimate calamity. Quite an opposite possibility will now be examined. (All values will be expressed in 1970 dollars.)

Suppose that the 1970 pattern of our good life were maintained continuously as an average, but the annual growth of the GNP were reduced to 2.5%. First raise the question: If the U.S. citizen could be satisfied with our 1970 level of individual and government spending per capita for all the usual purposes, could we really at this slow rate of growth provide the funds for solving such problems as energy supply and pollution control and better living for the underprivileged in our country and yet have a surplus for new ventures at home plus generous

Future Billions for New U.S. Projects

aid for the malnourished and the developing countries? This is an important question for the balance of the twentieth century. In the following paragraphs this speculative example is examined to focus attention on possible answers to the question just stated.

First we must examine the 1970 standard. Although there was inflation in 1970, and the stock market recorded a small business recession, the year was for the average citizen the best that had ever been experienced, with personal consumption twice what it was in 1950 (Table 33). Personal consumption per capita increased another 12%, however, before it leveled off in late 1973. Will postrecession citizens be satisfied with a 1970 living standard, or will they demand a 1973 standard or better? There is danger that once the postrecession recovery gets under way, most people will expect a return to a growth rate of individual income in the 5% range. If this is attained, another major adjustment can probably be expected.

The entire world insists, and we usually admit, that U.S. living standards are excellent, and economic indicators such as the GNP per capita place us at the top. Many adjustments are needed to cure inequalities, but more money for everybody is not the cure. Personal consumption spending in 1970 was $3015 per capita, and another $1755 per capita remained from the total GNP of $4770 per capita to meet all other expenditures. If our postrecession economy could return to these levels, in equivalent 1970 dollars, the fact is that, even with only 2.5% annual growth in the GNP, large sums would be available every year for new projects.

The assumption of a constant average level of spending by each individual, at 1970 rates, but with a 2.5% annual increase in the GNP, is built into Table 31.[47] The growth rate is expected to average 2.5% over each of the periods indicated, but admittedly the 1980 values in the table might be questioned because of the mid-1970s recession. But unless the recession lasts much longer than anyone predicts, the 1980 level of a $1250 billion GNP will be attained because the 1970–1973 increase makes up the 1974–1976 loss. After the recovery, the slow growth assumed is only about one-half the rate of the 1960–1973 period.

With population growing at about 1% per year, the total expenditures for usual purposes increase accordingly, but the normal standard of living remains at $4770 per year per capita. All excess funds would be used for new projects. For the year 1985 these surplus funds would amount to $285 billion, and by the year 2000 the new money would be

---

[47] Table 31. Future Billions for New Projects: New Funds Available for a 1970 Style of Life.

over $700 billion (in 1970 dollars). Table 31 indicates that, even with this assumed slow growth of 2.5% per year, the annual per capita expenditure for new projects in the year 2000 would approach the annual total available for all personal consumption and would be well above 1970 outlays for all purposes other than personal consumption.

Admittedly this is a highly oversimplified picture, and it could not be attained and continued for an indefinite period. Perhaps our citizens would not even consent to a constant average living standard no matter how high it was. But in view of our great material advantage over every other country in the world, even in 1970 (Table 1), it does not seem to involve a great sacrifice, because the idea of a constant average standard of living does not in any respect involve the leveling of individual opportunity or standards. It would not change the competitive private enterprise system or interfere with profits at the usual rates. With some exceptions for domestic relief and foreign aid programs, all products and services representing the cumulative surplus would be furnished by private enterprise to the same extent as in the past, in part by government contracts, as in space and defense programs in the past.

A reform modification of the tax system would be called for, however, designed to prevent escalation of consumer and public spending beyond the agreed growth rate. This would call for periodic evaluation of the trends, annually or oftener, with tax rate adjustments at intervals. Setting up such a program would require time for intensive public education and discussion. New social improvements, such as health and education programs, would constitute new projects to be financed from the 2.5% surplus. Capital improvements would continue, but a lower fraction of the GNP should be required for this lower rate of growth (about one-half the 1950–1973 rate), thus requiring a somewhat lower fraction of the GNP than the approximately 16% formerly used (Table 33).

While the results shown in Table 31 indicate that slower growth is not necessarily a disaster, they are obviously based on a change in public attitude that could not be quickly attained. The large amounts shown for new projects are based on the assumption that the U.S. population could be persuaded to accept a rate of per capita personal spending that remained at the real level existing in 1970 (constant dollars). This amounts to a redistribution of the benefits of growth. Instead of these benefits being used for immediate-consumption spending, they would be diverted to specific programs for building up the nation and the good life it offers.

An educational program of huge proportions would be necessary to gain acceptance of such an idea, but a sufficient nucleus of enthusiasts could certainly sell its advantages to a great many people. Decision by a bare majority to undertake the change would probably be insufficient, and no more successful than some other reform attempts (e.g., Prohibition, 1920–1933).

The possibilities of national accomplishment and of personal opportunity under the assumed change would certainly be appealing if a large and increasing "new projects fund" were available. Any planned diversion from personal direct consumption increases would doubtless be resisted by many, and the means for accomplishing it would be resisted by many more. Nevertheless this speculative exercise suggests a highly beneficial alternative to the limitless upward spiral of private and public spending for current consumption, some of which is of marginal value, to say the least.

Chapter Seven

U.S. Economic Expansion, 1970–2000

A. PAST PERFORMANCE AND GROWTH OF THE U.S. ECONOMIC SYSTEM

The grand plans of the futurists can be carried out only if the U.S. GNP continues to grow and is then reallocated to meet the new demands. Capital for the development of additional energy sources and for increased production of scarce metals, and funds for any and all aid for the malnourished and for economic assistance to less developed countries, must come from the total product of the country, the GNP.[48]

Can we produce enough to meet all these new demands? Actually, such a question cannot be tolerated. The demands *must* be met and can be met. Realistic estimates must be made of the funds required and of the reasonable limits of funds available; then allocations will be necessary. We already have the world's most successful economic system,

---

[48] The GNP is reported annually and analyzed in detail by the Department of Commerce (see *Survey of Current Business*). The GNP constitutes a summary of all goods and services produced during the year, plus additions to the total of structures and facilities, by capital investment. It includes all foreign aid, net exports over imports, all defense purchases, and government expenditures and services. It is computed separately on two bases, income and expenditure. It is notable that, while dollar quantities have consistently increased since 1950, by growth and inflation, the percentage distributions have changed very slowly, as is shown in Table 35.

and it should be changed as little as possible; but we now realize that past ideals of limitless growth and continuing increase in our living standards are due for some modification.

The great accomplishments and prosperity of the last 25 years show up clearly when the GNP and its percentage distribution are examined (Tables 33 to 35), but some future limitations are apparent. What are the practical limits of funds, manpower, organization, management? This chapter is intended to help the reader form some answers. We must start from where the economic system stands today, and this calls for facts about the immediate past. After reviewing the new demands, estimates must be made of future funding possibilities. Specialists in each field declare their dollar needs, but some of them call for capital and manpower far beyond what can reasonably be expected. Allocations will be required.

As background information for future planning, the past performance and recent expansion of the U.S. economic system and some comparisons of 1973 with 1950 are informative. While population and the labor force increased about 40% in this 23-year period, the real GNP increased to 270% of its 1950 level during that period. Expressed in average annual rates of growth, the GNP in constant dollars grew at the rate of 4.4%, while population increased at a rate of slightly less than 1.5%(Table 32).[49] Although the GNP per capita thus increased 3% per year, the percentage of the GNP representing personal consumption actually decreased from 67 to 62.4% of the total GNP, as shown in Table 33.[50] This means that other expenditures must have increased much more rapidly than personal consumption. The relative rate of domestic investment was not increasing so fast, nor were our net exports increasing materially. The rate of investment actually declined, and the net exports averaged less than 1% of the GNP. Government spending, however, did increase, from 13.3 to 21.5% of the GNP. The highest rate of increase was in expenditures by state and local governments (Table 33). While total personal saving in constant dollars in 1973 was two and one-half times what it was in 1950, it still accounted for only about one-fourth of private domestic investment (Table 32). The capital for business expansion always came largely from business itself.

Table 34 gives a breakdown of the GNP on the basis of income rather than outgo, indicating that the percentage representing personal income was remarkably constant at about 80% of the GNP throughout the

---

[49] Table 32. Some U.S. Economic Comparisons, 1973 versus 1950: Economic Activities in Current and Constant Dollars.
[50] Table 33. Distribution of the GNP, 1950–1973: U.S. Statistics, Outgo Basis.

1950–1973 period.[51] The percentage of personal income going to taxes increased, however. Thus the percentage representing disposable income decreased (Table 34), and this decrease corresponds to personal consumption (the first item in the breakdown in Table 33, the outgo basis).

Table 34 lists corporate profits and dividends as a percentage of the GNP, and these both declined about one-third during the 1950–1973 period. This trend substantially contradicts the statements of agitators during the 1960s to the effect that business and industry were taking a larger share in the form of profits. After all, profits are the main source of capital for improvements that will increase productivity, making pay raises possible. This decline in profits is again shown in the outgo analysis by the reduction in the percentage of the GNP going to private domestic investment (Table 33). Personal saving increased less rapidly than the GNP between 1950 and 1973 (Table 32).

Incidentally, other industrial nations have recently been much ahead of the United States in improving plants and facilities and in the research and development of new products. The percentage of their GNP for these purposes was much higher than ours, and it was increasing rather than declining. We were even helping many of them in this respect with our postwar aid to Germany and Japan. These trends have left us in an inferior position in several industries, with obsolescent equipment and less advanced products, hence a poorer competitive position.

A more detailed percentage breakdown of the GNP in Table 35 allows some comparisons within the three segments of the economy, that is, industry, government, and consumers.[52] In general, the rates of capital investment remained nearly a constant fraction of the GNP, although in the utility industry they increased about 50%. The most conspicuous increase among federal programs was social security. It should be noted that the government calls this a "transfer" item, because disbursements are designed to balance collections. This is somewhat unfortunate, because it gives the impression that one segment of the population is supporting another, whereas a retiree who has paid social security tax most of his or her life is of the opinion that this is an insurance or retirement annuity program.

Table 35 indicates that exports of US. goods have represented an increasing percentage of all goods and services. But since imports have

[51] Table 34. Distribution of the GNP, 1950–1973: U.S. Statistics, Income Basis.
[52] Table 35. Percentages of the GNP, 1950–1973: Details of Incomes, Expenditures, and Investments.

increased at almost the same rate, the net export drain on the economy has averaged less than one-half of 1% per year. Consumer spending has been decreasing, but the fraction for durable goods has remained almost constant, while expenditures for services have increased at the expense of the purchase of nondurables.

All statistics on the distribution of the GNP are actually less accurate than the tabular data, to two or three significant figures, indicate. Government analysts admit this, and values of statistical discrepancies are included in their tables, as indicated by the last item in Table 35.

Additional information on the performance of the U.S. corporate system is presented in Table 36.[53] This table includes some averages of the samples at 5-year intervals from 1950 to 1970, that is, averages of the five individual years. To aid in interpretation, certain ratios are shown, such as corporate taxes with respect to profits and dividends, ratio of profits to gross income, and expenditures for new plant and equipment as related to profits and dividends.

B. THE LIMITS OF CAPITAL EXPANSION AND OF FOREIGN AID

There is every indication that a combination of great demands for capital investment at home and for foreign aid for other nations will be a major characteristic of the postrecession economy for the balance of the century.

It is helpful to review the specific purposes for which new funds are urgently needed:

1. Capital to make possible large increases in coal mining and usage and increases in U.S. production of oil, fuel gases, and electric power, and to provide new energy sources and energy conversions, all with accompanying provisions for adequate pollution control.

2. Capital for facilities to increase greatly the production of fertilizers and other farm chemicals, both for our own farm use and for export, largely to the less developed nations.

3. Capital for bringing additional land under cultivation. Large increases in food production are required to prevent imminent starvation in South Asia, Africa, and Latin America. This will require much more land than is now under cultivation in the United States, since we are now the

---

[53] Table 36. Performance of the U.S. Corporate System, 1950–1973: Total Profits, Dividends, and Depreciation Allowances.

main food-exporting nation in the world. New funds must be available for land clearing, irrigation, drainage, and the more costly farming on this less desirable land.

4.   Financing of adequate production of U.S. consumer goods sufficient for early improvement of living standards of the less fortunate in our own country.

5.   Large additional contributions and long-term loans for economic develoment in the less developed countries.

6.   Major participation in international action to subsidize underdeveloped countries, reduce malnutrition, instruct in technology and know-how, and meet emergencies and financial crises.

These needs *must* be met. By what means and how rapidly can they be accomplished?

An initial step is to examine past experience in financing new projects and compare it with what these new challenges will demand. If past methods are inadequate, what new measures are necessary?

Several of the objectives on the above list are beyond the means of private initiative. Federal government organization and assistance will be required, especially in research and planning, and later in determining priorities and regulating operations. Sudden increases beyond established financing patterns cannot be expected, so it becomes important to examine the magnitudes of past funding accomplishments, both in private and government sectors. Expressing these in terms of percentages of the GNP helps to avoid unreal expectations.

Before attempting dollar estimates, it is well to review the distinctions between government functions and those of private corporations under the American system. The U.S. government has seldom engaged in construction, mining, manufacturing, or agriculture, except to meet certain military needs or for flood control or land reclamation, and U.S. citizens are not likely to authorize major activities of this kind in the future. They consider these the province of private industry. The efforts of Congress and of the executive departments have appropriately been reserved for policy planning and execution, and for providing subsidies and assistance where competition and the profit motive are inadequate. When the government undertakes business operations, these functions are neglected. The democratic process is too slow for everyday business decisions.

Capital for new power plants, oil refineries, gas wells, shale oil plants, or fertilizer plants must be procured in the usual way from savings and investments by private individuals and corporations. Once the facilities are built, the government may become a major customer and may even subsidize the operation directly or indirectly. The bulk of all capital funds comes from personal savings and corporate depreciation allowances and profits. It is a common practice of American corporations to retain over 50% of yearly profits for company growth rather than paying out these amounts as dividends to stockholders (see Table 35). For the new projects listed above, however, the government will very likely play a large role in causing some of these capital funds to become available.

It would be foolhardy to attempt a detailed prediction of the total costs of financing each of the six objectives. It cannot be expected that these projects will proceed smoothly, without great ups and downs from year to year. Predictions will be examined only on the basis of averages, with emphasis on some samples in the energy field. There are really only three general fields involved: industrial capacity, agricultural capacity, and international participation in helping people who are in much less fortunate circumstances than our own.

1. *Estimated Funds Available for Private Investment, 1970–2000*

Accomplishment of the six objectives listed above is a real necessity, and capital supply promises to be a major limitation. How fast can we proceed? Planners are likely to be too optimistic and set unrealistic target dates. Prior to the time of President Nixon's proposal for Project Independence the year 1980 was a common target date, but it has since beeen recognized that it will be at least 1985 before the success of the several proposals can be ensured. In the following analysis of the prospects, the year 1985 has been selected in an effort to arrive at a realistic time schedule. Estimates will be expressed in percentages of the GNP, with monetary sums in constant, 1970 dollars. Among the immediate questions are: Is it possible to increase the rate of saving for investment? How do our rates of saving compare with those in other countries? Is increased capital investment really important?

As to the last question, there is already an indication that our labor force capacity exceeds our capital capacity. A period of inadequate capital investment eventually results in a condition of not enough jobs, because every added worker calls for added equipment. There have been several reasons why adequate investment in new plants and equipment has been difficult for many years. Lowered corporate profits in percentage of the GNP, especially in the 1970s, have resulted in lesser amounts being retained for investment in business (Table 35).

Diversion of capital to foreign plants, and also investments for non-production purposes such as those required by the EPA and the OSHA, have reduced the amounts left for production equipment and moderniza-tion, and for research and new product development. Inflation has in-creased replacement costs, and government borrowing has raised inter-est rates, making it unattractive to finance capital purchases by loans. Military and space programs have encouraged private capital invest-ment for nonconsumer purposes. Productivity per work-hour has been retarded because plants failed to modernize, and foreign competition has therefore increased in many lines.

As a comparison with other countries, the fact is that Japan's capital investments during 1960–1973 were twice as large as ours, as a per-centage of the GNP. In this respect we are in the low company, below the United Kingdom and Italy. West Germany and France invest much more. We are in a like position with respect to our total efforts in research and development. Investments in plants for new and improved products can only follow productive research programs, and several industrial coun-tries are way ahead of us in the percentage of yearly expenditures devoted to research and development. Almost none of the highly trained workers and scientists in Japan and West Germany are diverted to military, space, or foreign-plant production, and almost none of their capital is so diverted.

In 1973 U.S. private domestic investment was $202 billion (Table 33). Investments that resulted in jobs and products for the American worker were reduced by diversions of U.S. capital abroad and by in-vestments for nonproduction purposes. Some products, especially mili-tary armaments, do not contribute to economic goods for the consumer.

For many years U.S. capital investment abroad has been substan-tial. While foreign production and sales may be highly profitable to the corporations involved, these investments provide neither jobs nor goods at home. For each employed U.S. worker a corporation must acquire plant and equipment capital of about $10,000 for simple manufacturing or commercial lines, and up to $200,000 or more for certain heavy goods manufacturing, or in gas or electrical utilities (see Chapter 3, Section E). The heavy wave of criticism of U.S. multinational corporations is due in no small measure to their diversion of capital to foreign plants.

A recent capital study which included data on twenty multinational corporations operating, on the average, in twenty-five countries, showed an average of almost 40% of their assets were located abroad, as were 38% of their sales. These were large corporations with total assets well over $100 billion for the group. One-half of them were among the fifty largest corporations in the United States. Even this partial sample of over

$40 billion foreign investment would furnish the plant and the tools for a payroll of perhaps a million workers if used at home. (Our employed civilian labor force is less than 90 million.)

Future funds available for private investment will depend directly on the rate of growth of the GNP, assuming for the moment that there is little or no change in the 15.5% average fraction used. It is probable that this percentage will not be long maintained unless the earlier and higher percentage of corporate profits as a fraction of the GNP can be restored. This is the major nonpersonal source of savings for investment (Table 35). The public must be educated to appreciate that the function of profit is to build up the nation's productive plants. It does this both by paying dividends that encourage more capital investment, and by direct investment of the balance in new plants and improved equipment. Misunderstanding of these facts about profit has caused many a battle in legislative halls. Profits are even more important in times when common stock prices are low and new equity financing is thereby made more difficult. Consumers share the ignorance and mount campaigns to reduce corporate profits. They think in terms of indefinite growth of consumption rather than in terms of making it possible to produce the goods they need or want. The time has come for a reduction in the advertising designed to create more unnecessary consumption. Of much more importance are energy, environment, and food programs, and also greater efforts toward world stability through helping the developing nations. Ralph Nader seldom mentions that the very employment and income of the consumer depend on a reasonable level of profits to buy the new equipment that makes more jobs possible. For several years the annual influx of new workers has been about 2.4%.

The public sees mainly the larger and larger dollar numbers representing the profits of corporations. They overlook the fact that part of this is inflation and that the growing population and its better quality of life can only be served if profits are increased to provide ample and improved additions to corporate plants and equipment. Profits must be viewed in terms of the total economy, as a percentage of the GNP, and this percentage has been decreasing. (There are better indicators than the GNP, but accountants and statisticians do not use them.)

As to the future total GNP from which the 15.5% might be available for private investment, we have selected a 2.5% annual (slow) growth as the most likely and desirable rate, but there have been many predictions of a higher rate of real GNP increase. President Ford, in his 1976 budget proposal, admitted there was negative growth of the GNP in 1974 and 1975, but predicted a 6.5% annual growth for 1978 through 1980.

From five authoritative studies of future capital requirements, including one made for the New York Stock Exchange, the average value predicted for future annual growth of the real GNP was 4.1% (terminal year not uniform). Incidentally, none of these five studies foresaw any large investment gap or capital shortage at this rate of growth, provided the average rate of private investment, 15.5% of the GNP, could be maintained, even though this percentage is the smallest reported for any major non-Communist industrial nation.

Persistence of a business recession has contradicted many earlier authoritative estimates of how much investment capital will be available up to 1985 and what can be accomplished with it. Many forecasts based on an average annual increase in the GNP greater than 3% predicted no great shortage of new capital.

Inflation has made dollar estimates hard to compare. It is already difficult to think in 1967 dollars or in 1970 dollars, the two levels commonly used for constant-dollar comparisons. However, complete published statistics always lag behind current experience, and this book uses 1970 dollars for comparisons. Wherever possible, the amounts are also expressed in percentages of the GNP, because these percentages are remarkably stable (Table 35). It should be remembered that the average GNP, an arithmetical average of the real GNP over a period of years, is used here for convenience only. When the economy is growing, the real GNP in constant dollars increases every year.

The slow growth rates we have selected for Table 19 appear in the first column of Table 37.[54] These are 2.5% average annual growth of both the GNP and total energy and 4.5% annual growth of electric energy from utilities. Two higher rates of growth are also tabulated in each case. Growth rates in the last column are much closer to the average historical rates of growth during 1950–1973.

The recommended estimate of growth of the GNP is from $977 billion in 1970 to $1415 billion in 1985, and to $2050 billion in the year 2000 (constant dollars). Over the 1970–1985 period, the average amount available yearly for private domestic investment, at 15.5% of the GNP, is about $185 billion.

The only use for these figures is to provide, from past experience, a "ball park estimate" of the capital available per year for private investment, from 1970 to 1985, to indicate the probable distribution of this capital, and to compare it with some of the estimated needs for capital during this period. During the years 1976–1980 growth will probably be greater, but the additional capital will be used for catching up on

[54] Table 37. Average Growth of GNP and Energy Supplies, 1970–1985: Monetary Amounts in Constant 1970 Dollars.

investments that were deferred during 1974–1975. Actual expenditures for new projects will be small at first, increasing from year to year.

2. Investment Funds Required for the Program, 1970–1985

The final purpose of these estimates is to determine whether the accomplishment of the six objectives listed at the beginning of this section can be substantially ensured by 1985. In order to do this, the probable capital required for the program must be compared with the estimated funds available. If the accomplishment looks impossible, there are always three alternatives—extend the duration of the program, cut the size of the program, or obtain more funds.

With respect to the third alternative, it *should* be possible to influence the employed citizen to save more and invest in the future. Reducing inflation would help stimulate the incentive for saving. The economic standard of the individual doubled from 1950 to 1973, since the real GNP increased 2.87 times while the population growth was only 1.38 times. Consumer income remained very close to 80% of the GNP for 25 years, but the consumer saved no greater percentage of his or her income than in 1950.

The other source of investment capital, corporate profits retained for investment, has been diminishing with respect to the GNP, and this trend must be reversed. Increased rates of capital accumulation by corporations can be promoted by incentives set up by the government. Among these incentives would be higher maximum allowances for depreciation, investment tax credits, lower taxes on capital gains, and allowing moderate dividends to be designated a corporate expense. Government borrowing should be minimized.

If private investment capital could be increased, even to 20% of the GNP, which is below that of several other nations, including Japan and West Germany, large additional funds would be available yearly for new projects. At the 2.5% growth rate of the GNP, the difference in the year 1985 would be $64 billion.

The rate of future growth is basic to all estimates of capital funds available or needed. A main thesis of this book is that the rates of growth experienced in the United States since 1950 cannot be expected or even allowed to continue for the next 25 years, both because of the limits of the resources and capacities of our planet and because of the human crowding, strife, and unhappiness that would result. While this slow growth conclusion is gradually being admitted, and it is supported by more and more authorities who have made adequate studies, there remains a giant educational assignment to modify our country's philosophy which has long been based on maximum possible rates of growth.

One of the largest future demands for capital will be for enlarging the supply of electric power. Several estimates of the costs of this expansion have been made, and some have already been mentioned (Chapter 3, Section B). In the three-volume report "The U.S. Energy Problem," made for the NSF in 1971 (footnote 16), fifty-six earlier forecasts were analyzed. The report then predicted an annual growth rate of 2.8% for total energy usage for the period 1970–2040. The same report chose an average annual growth rate of electric generating capacity of 5.02%. In 1972 the FPC predicted a 5.9% growth rate of electric power to the year 2000. In 1974 several studies favored continuance of the recent historical growth rate of about 7% per year. A much quoted estimate in 1974–1975 was $500 billion for expansion of the electric power industry before 1985. If this investment were averaged for the 15-year period 1970–1985, it would call for $33 billion per year.

Our preferred estimate of growth of electric power is 4.5% per year, as shown in the first column of Table 37. At this rate of growth, the power generated in 1970, 1.53×10^{12} killowatthours, could increase to 2.96×10^{12} kilowatthours in 1985. Starting from a capacity of 341,000 megawatts in 1970, it would be necessary to finance another 319,000 megawatts during the 15 years. Translating this increase into (constant) dollars required would call for a knowledge of the design mix as regards kinds of fuel and percentages of nuclear installations of various kinds. For costs of generating stations only, a round figure of $500 per kilowatt has been used, giving a total cost for the new capacity of about $160 billion, or an average of $10.7 billion per year over the 15 years. Costs of distribution and sale of this increased output must be added, plus the costs of any unusual environmental protection. This could double the total cost.

Much more is involved in this expansion than the cost of generating plants and distribution systems. Fuel consumption, fossil or nuclear, is doubled, thus requiring great expansion in the facilities for mining, transportation, and delivery of the fuel. The unit costs of fuel will be higher, and the costs of converting gas- and oil-fired stations to coal will be substantial. The water used, and the pollution generated by these large increases in plant capacity, present major difficulties, especially as regards the siting of plants. The long delays for site approval that have recently become common must somehow be avoided if plant capacity is to be completed at the necessary average rate of 20 million kilowatts or more per year. The plant siting difficulties and the facilities for water supply and pollution control will increase costs, and inflation always adds to the difficulties of financing. In addition, the entire program will compete with all other construction activities for scarce manpower for the engineering, technical, and skilled-labor assignments.

For the expansion of the *entire* energy industry, one overall figure which has been quoted by such authorities as the chairman of the board of the General Electric Company, and which has received wide publicity, is a total capital cost of about $3300 billion. Presumably this must cover new coal mines, new oil and gas wells, large water supply and cooling facilities, new refineries, synthetic gas and oil plants using coal or shale, development of new and better engines, electric cars, and trucks, large additions to mass transit, development and commercialization of breeder reactors, all pollution control equipment associated with the use of energy, storage facilities for nuclear wastes, and much more. Whatever the program, $220 billion a year is a very large investment (see next section).

The above estimates of future capital costs are a small sample from many careful studies of the investment required to build up our energy supplies. This is only the first of the six items listed at the start of this section, and the examination has been based on the 1970–1985 period only. What about the other five items?

Present purposes do not allow a detailed attempt to answer this question, but a mere reading of the other five items listing purposes for which funds are needed indicates that only very strict financial planning can satisfy these interdependent demands. Fertilizer plants, some of them probably located near distant oil fields, will need billions of dollars per year (item 2). Increased agricultural output on more land (item 3) cannot be accomplished at today's unit prices. The cost of land preparation and water supplies alone will add billions of dollars. Everything depends first on rates of growth of populations, but energy supply is the most immediate problem of those listed. Even a crude analysis of costs in the energy field alone produces a conviction that the overall funding operations are in no sense an isolated assignment for bankers and politicians. A holistic approach is absolutely necessary. Urgings of single-purpose enthusiasts and of disciplinary specialists must receive due consideration. In the following section, dealing with the adequacy of available capital for investment, undue emphasis on energy is not intended. The complexities and significance of funding to deal with food shortages is fully admitted and of equal import. Environmental control and the needs of the developing nations are included. All must be served from the same reservoir of funds provided by the U.S. taxpayer.

3. Will There Be a Capital Shortage?

The average of all private U.S. capital funds each year during 1970–1985, at the preferred annual growth rate of 2.5%, is about $185 billion (Table 37, line 5). But this amount is for *all* capital investment. How much of the $185 billion will be available for growth in energy and in agriculture? The answer can be estimated from statistics on past investments, since sudden changes are very unlikely. The next question concerns the distribution of available funds.

Residential construction averaged about 30.6% of all private investment during 1950–1973 and, according to *Construction Reports*, other private building construction was another 15.2% of private investment. This leaves 54.2% of private investment for all other purposes. Included in the latter is a small amount of yearly addition to inventories when the economy is growing (Table 33). Investment in farm machinery and vehicles is about 3.5% of total private investment, reducing the balance to a 50.7%, or $93.8 billion, average per year for 1970–1985 at the preferred 2.5% growth rate of the GNP. Most of this is used by industry and business.

An approximate distribution of the $93.8 billion can be learned from "Business Expenditures for New Plant and Equipment," as reported by the Bureau of Economic Analysis (BEA), in *Survey of Current Business*. From this source, for the 5 years 1969–1973, it is learned that the industries *least* involved in the production of energy accounted for an average of 59.6% of the capital investments of all industries (Table 38, Item 14).[55] In this nonenergy classification the author has included all the business groups of items 3 to 12 in Table 38, namely, food, paper, chemical, rubber, and other nondurable goods manufacturers, transportation and communications industries, and durable goods manufacturers using nonmetallic raw materials. Included also (item 12) are construction, trade, service, finance, and insurance. (Entirely excluded from Table 38 are professions, institutions, and real estate firms.)

Our main interest at this point is in the future growth of the energy industries. Subtracting from the previous balance of $93.8 billion per year, for 1970–1985, the 59.6% used by the industries just named (1969–1973) leaves roughly $38 billion per year. This is an approximation of the amount that would be available for fuel production and preparation, the energy utilities, all mining, petroleum and natural gas, and all metal goods manufacturing including the automotive industries.

---

[55] Table 38. Capital Investments by U.S. Nonenergy Industries, 1969–1974: New Plants and Equipment (billions of current dollars).

Thus the $38 billion includes, in addition to capital for fuel and energy production and utilities, a sizable amount for the metals and machine industries. From past experience the capital required for metal goods manufacturing in this case would be close to $12 billion per year.

Private utilities use about 1.25% of the GNP for their capital needs, according to Table 35. For the estimated average for 1970–1985, an approximately $1200 billion annual GNP, the utilities would use about $15 billion per year, of which the electric utilities would use over $12 billion.

The purpose of these several calculations is to obtain approximate estimates of the amounts of capital that would be available on an average yearly basis for the period 1970–1985 if the rate of growth of the GNP were 2.5% per year. The results can be summarized as follows:

1. Out of an average $1200 billion GNP, $94 billion a year will be available for private capital investment in industry.
2. Capital for nonenergy industries is estimated at about 60% of the $94 billion, leaving $38 billion for direct energy-related investment.
3. Of the $38 billion, $15 billion will be needed for utilities, leaving a $23 billion balance about equally divided between the fuel-supplying industries and the metal-manufacturing industries.

Now the remaining question is: Will this investment budget provide for the capital needs already outlined? Since these estimates are based on past experience, with rapid growth, and the 2.5% average future growth will be slower than that before 1974, capital requirements should actually be lower.

Before trying to compare the capital funds that may thus be available in the future with the demands on these funds, it should be emphasized that competition for capital will be intense. Due attention should be given to each need for capital funds. If the funds available in a given year are inadequate to cover projected expenditures, there are three ways to match the two: (1) Scale down the expenditures for each project; (2) increase the total funds, as by transfer from other uses; or (3) extend the time for completion of projects. It is desirable to work for changes in the distribution of the GNP that will increase the percentage available for the six objectives listed at the start of this section.

Consider as a prime example the electric power utilities and their suppliers. Electrical utility companies are estimated from past experience to have available to invest an average of about $11 billion per year for the 1970–1985 period. This should be a high limit, because it is based on their share when they were growing much more rapidly. This amount is only very slightly more than the $10.7 billion per year shown in Table 37 (first column, last item), and the latter amount was for building generating stations only and did not include additions to distribution systems or office expansions. Twice that amount would probably be needed, hence a large capital shortage is indicated.

The previously mentioned $500 billion intended to cover *all* capital investment for expansion of the electrical utility industry would amount to $33 billion per year to 1985. Since only $38 billion is to be available for *all* direct energy-related investment, a further large shortage of capital is evident.

The $33 billion figure may be presumed to include most of the $100 billion total proposed by President Ford for a comprehensive research and development program in the energy field, since he expects most of the funds to come from the private sector. The government may finance an amount equivalent to several years of the earlier space program. (The latter cost the government about $7 billion in its maximum year.)

The outside estimate of $3300 billion already indicated for the necessary expansion of *all branches of the energy industry,* including investments for environmental control, would call for an average of $220 billion a year from 1970 to 1985, and even much more from 1975 to 1985. This estimate has been quoted by many authorities and, if it is at all realistic, a major capital shortage must certainly be expected. The summary of amounts of capital probably available to 1985 (see above) indicates less than one-half as much as this, $94 billion, *for all industry and business,* and only $38 billion per year for all energy-related investment.

Almost every plan for economic or social improvement calls for capital investment. In addition to large outlays for energy-supply facilities and environmental protection, there must be a spurt in capital investments in all industry to make up for the lag during the recession. When the unemployed are put back to useful work, investments in tools and work place equipment will be necessary. Even at a minimum rate of $4000 per person (used above), this will amount to about $4 billion for each one million employees. In addition, the normal growth of the work force is nearly one million per year.

In view of such figures it is time to conclude that investment capital is going to be in very short supply indeed for the next decade or two. All information derived from our historical pattern indicates that the several objectives already discussed cannot be met by 1985, because of shortages of capital.

Such conclusions are based on the *slow* future growth that this author thinks is an imperative, dictated by shortages, by crowding and environmental problems, by slower growth of the U.S. population, and by international pressures. The authoritative studies that see no great shortage of capital in the near future are based on the assumption of a return to rapid growth. Typical is the 1975 Brookings Institution report compiled in 1973–1974, which concludes that "with normal growth and without unusual sacrifices the economy will be able to meet the capital demands that can reasonably be projected for the remainder of the decade." (Bosworth, Duesenberry, and Carron, *Capital Needs in the Seventies*, Brookings Institution, 1975). This study assumes that the GNP will grow at the average rate of 4.3% per year for 1970–1980, which is slightly above the 1960–1972 rate. However, the study was made before the negative growth rates of 1974 and 1975 were reported.

Growth rate makes a great difference, as illustrated by the following, computed for specific years. If the 4.2% rate of growth in domestic product that was the average for 1960–1972 is replaced by an average growth rate of 2.5% as per Table 19, there will be about $36 billion less capital available in the single year 1980 and about $208 billion less in the single year 2000. This is based on average growth at the two percentages, starting in 1970, with capital at 15.5% of the GNP.

It is our conclusion that all possible means for increasing the supply of capital for private investment must be applied. It is important, even necessary, that personal saving and reinvested profits be increased. This can be encouraged by changes in the tax laws, accompanied by vigorous educational programs to correct today's misapprehensions on the part of the public with respect to the functions of profits and investments. We cannot hope to attain Japan's rate of capital formation, but we should be able to match West Germany or France. We must actually expect to extend the period of newly planned development efforts and investment programs to 1990 or beyond, because the objectives cannot be attained by 1985.

During the 3-year period 1974–1976, the optimism of the energy industries as to their total growth by 1980 was considerably reduced. The 1974–1975 depression proved more severe than anticipated, and the 1975–1976 recovery much slower. Estimates of total capital needed by the energy industries in the 1975–1985 decade began to be revised downward, especially after the several congressional actions that withdrew the oil depletion allowance, retained price controls on both oil and natural gas, and failed to support the president's recommendations for heavy funding of energy research and development. (See the ERDA plans, "Energy Choices for the Future," ER 1.11-ERDA 48, Vols. 1 and 2, 1975).

These overall developments forced the conclusion already indicated, that energy objectives cannot be attained by 1985. This time the reason is not that such vast sums will be needed that the capital shortage will be acute, but rather that these great sums will not be needed. The people through Congress, and the economic system through its sluggish recovery, have decreed that the capital needs for supplying energy will not even approach previous estimates of up to $220 billion per year.

This new viewpoint would relieve the great urgency of increasing private capital formation. It remains to be seen which of the energy futures is the more accurate, but either viewpoint tends to discredit the likelihood of vast expansion in electric power combined with ample domestic supplies of oil and gas based on energy independence rather than on increasing imports. This points to a more modest energy future in which the need for conservation will be paramount.

4. Emergency Foreign Aid and International Obligations

What is the size and nature of world food relief needs? As indicated in Chapter 4, food supplies are a long-term problem, and the situation is growing worse because in most of the developing countries the population is growing faster than the food supply can possibly be increased. More than one-half of the human race is malnourished, and total food stocks are now very low; so that after recent crop failures in many areas, starvation is increasing at a sickening rate. Some cooperative method for international storage and shipping of grain reserves is badly needed.

For the hundreds of millions who must spend three-fourths or more of their income and effort for food the only possible result of a doubling of food prices is less to eat. In most cases their diet is already inadequate.

Recent food price increases within the United States itself have caused much dissatisfaction, and this fact combined with the attitudes and pronouncements of the small-nation majority in the UN have made it much more difficult to obtain a favorable response to emergency foreign aid needs, as shown by the sharp debates and turndowns in Congress. Higher prices also mean much less food per dollar of food aid funds.

North America has only recently become the major food-exporting area in the world. U.S. exports of food as crop or animal products have recently been about $12 billion per year (see Table 21). The fraction of U.S. cropland now used for export products is almost double what it was in 1950–1955. The very fact of this great increase in food production, with no corresponding increase in the number of farmers, gives us great opportunities as well as moral obligations. There are vast areas in the world where our agricultural know-how could produce great results. In many cases the exporting of our methods and our capital could even be done at a profit, especially in large-scale farming projects on land not now being farmed, in South America in particular. Such projects would require large amounts of fertilizer and often irrigation too, but these are actually further opportunities. In many locations in the world there are known to be large untapped sources of natural gas, while in many others the gas is flared off during oil production. These locations may be distant from markets for gas fuel, but they are choice locations for fertilizer plants which could use the processes for making fertilizer that have been so fully developed in Texas and elsewhere. We could provide such plants. The methods of water conservation and irrigation used in our arid states are much needed in other arid countries, and we could export these skills also. Projects like this would provide much needed employment for local populations and would show their leaders what modern technology can accomplish.

An American opportunity not often mentioned is the solution of our own diet problems in this process of helping others. Our overeating and overweight, with a preference for foods high in animal protein and fats, increase the likelihood of heart disease and other maladies. The practical, worldwide result of our habits is that we use nearly a ton of grain each year per capita, with almost 90% of it going to the feeding of animals. Some changes in our diet and in our animal population would free a lot of grain for the poor nations where only 400 pounds of grain per capita is used each year. We would save money in the process, because meat is much more expensive than vegetable protein such as is available from our vast crops of soybeans and other legumes.

Tentative general conclusions on the food export and relief situation are that U.S. food relief programs that have recently been in the $1 billion to $1.5 billion range must be greatly increased immediately, perhaps tripled to $4 billion plus, as soon as possible. These gifts and emergency subsidies will then approach the yearly costs of the space program of the 1960s. Our entire foreign aid program would be more than doubled, as compared with its level over the past 15 years. The higher costs of producing the additional food will also call for subsidies. Ordinary market prices would not be able to support great increases in the use of fertilizers by our farmers, plus the costs of preparation of the new land needed, namely, clearing, drainage, or irrigation costs. There will be diminishing returns from using more fertilizer on present cropland to increase yields, and on new lands of lower fertility.

The early UN goal for assisting the underdeveloped countries was 1% of the GNP from each industrialized country, or $10 billion for each

$1 trillion of the GNP. If the world shortage adds to the U.S. budget $2.5 billion per year for subsidizing crop expansion and another $2.5 billion for food relief, the total aid and subsidy, say $10 billion per year, will approach the UN 1% goal. There is no doubt about the ability of the United States to meet such a foreign aid program; but it will call for some changes in the American style of life, especially when added to the energy program and related environmental programs. After all, these amounts for our share in world development are only a small fraction of those indicated by the Second Report of the Club of Rome, as described in Chapter 8, Section B.

In the final analysis, the success of all food aid programs must depend on early progress in reducing the rate of population growth in nations where the needs for food are most acute. Can they be convinced that the eventual choice is birth control or starvation?

C. THE LIMITS OF MANPOWER

A long-term slowing of U.S. economic growth has been widely predicted, and was advocated in the previous chapters, based in part on limitations of natural resources and environment but indicated also by the expected slower growth of the population. There is no good reason why shifting the economic emphasis from quantity to quality should impair the good life, even if some life-style changes would be necessary, although it is always difficult to abandon comfortable old habits and adjust to new demands. The purpose of this section is to examine briefly some of the new demands on the work force with reference both to numbers and to vocational makeup, and thus to estimate what manpower limitations, if any, may be encountered in meeting these demands.

The measures already outlined for dealing with supplies of energy and food, and with environmental problems, call for great increases in capital facilities, for more general use of available, existing technology, and for much research and development in the near future. There must be extensive construction of new facilities for the extractive and manufacturing industries and for agriculture, with corresponding expansion of service facilities, including necessary transportation, communication, and power supply. Almost every process involved will require additions to water supply systems and to pollution control equipment. All these demands call for more manpower, much of it specialized.

Slower population growth is already affecting the schools, and it will soon reduce the large work force, aged sixteen to thirty-five, to which the country has become accustomed. The net annual increase in U.S. population first attained 2.5 million about 1950 and stayed above that figure for 15 years, but the additions are now a million less per year, in spite of a population 60 million larger than that in 1950. There annual increase in population has dropped from 1.8% to less than one-half that rate (see Table 39).[56] Birthrates, which were near 25 per 1000 population in the 1950–1960 decade, have dropped almost 40%. Since unwanted or unplanned births were reported as 42% for whites and 61% for blacks for the 1966–1970 period,[57] it is likely that birthrates will fall even more as contraceptive methods become more widely adopted. The number of new recruits in the labor force per year will soon begin to decrease and will continue to decrease after 1980.

The slower future growth of the U.S. population, estimated to average only 0.8% per year for the next 25 years (Table 19), will result in a decrease in annual housing starts and in the construction of shopping centers and other consumer facilities. Some of the displaced workers will then be available for the large construction programs already mentioned, in the energy industries and elsewhere. The very rapid recent growth of the service industries will be slowed, since the rate of growth of the total labor force will be decreasing. According to the U.S. Bureau of Economic Statistics, the employment in service occupations increased 94% from 1950 to 1973 (Table 39). The U.S. population increase for the same period was only 38%.

Based on such data as the above, it must be concluded that the work force available to meet the described future demands may be less than adequate. But from past experience, such as that with the space program, workers can probably be made available for most assignments, and this includes construction workers. However, since much research, development, and new design are involved, technical personnel will be in short supply, especially engineers and engineering technicians. Few of these will be released by slower growth of the housing, building, and consumer goods industries, because few are employed there now, hence most of the new recruits must be engineering graduates or technical trainees.

While enrollment in colleges has been steadily increasing and has doubled since the mid-1950s, and the percentage of high school graduates entering college has been increasing, this is not the case in en-

---

[56] Table 39. Growth of U.S. Manpower Since 1950: Trends That Will Affect the Future.
[57] "Population and the American Future," U.S. Commission on Population Growth and the American Future, 1970.

gineering and the sciences. There have been major increases in enrollments in law, medicine, most business fields, and the social sciences,
but engineering enrollment dropped rapidly after 1950 and has taken 25
years to approach the earlier output of over 40,000 bachelor's degrees
per year (Table 39). A much smaller percentage of entering college
students have been electing engineering and science.

The Bureau of Labor Statistics has estimated annual engineering job
openings for 1975–1985 will average 75,000 per year, and specialists in
engineering education say that up to 1980 the average supply from
colleges will be less than 40,000, since many graduates go into fields
other than technical engineering. While engineering colleges respond
quickly to demand, it takes about 6 years for an engineering education
and apprenticeship; and the added students are not yet in school, so the
demands of the late 1970s cannot be met by the colleges.

This means several things. Some projects will be delayed, but the
programs for new energy sources, and so on, are getting a very slow start
anyway. Junior engineering assignments will be taken by industry
trainees transferred from other jobs, and from the upgrading of engineering assistants and technicians. Nevertheless, it will be most difficult to
live up to an optimum schedule for research, development, pilot operations, and expansion of facilities. More time will be spent in cut-and-try
testing and improving of equipment, because of inadequate research
and design.

The limits imposed by manpower shortages will also be greatly
affected by political maneuvering. Technical training is expensive, and
engineering schools have been in a financial depression for some time.
There is little promise that the added expenses of increased enrollments
will be readily provided as they were by federal aid in the 1950s or the
research grants of the 1960s.

D.   THE ENVIRONMENTAL LIMITS

"Assurances are high that the present living generation in the U.S. will
never be severely penalized by pollution limitations. Timewise, any
widespread limitation to be imposed on the U.S. economy by the natural
environment is distant indeed."

Perhaps such controversial statements as the above are considered
a great disservice by residents of inner Los Angeles, struggling to be rid
of smog, or by the New Orleans Chamber of Commerce, fearful that
rumors about their city water supply will cancel the tourist business. But
this is just the point. Environmental problems to date are localized and, if

properly dealt with, can be temporary. This does not mean that we should discount the efforts of crusading ecologists trying to preserve the wilds of Alaska, nor that warnings are unimportant as regards the water shortage and the land pollution that could accompany major shale oil recovery in Wyoming. It means, rather, that we can count on the adaptive control of such projects by concerned people. Warned of the dangers, citizens can be depended on to protect the environment. There are many pessimistic views to the contrary, but history supports an optimistic conviction.

We are greatly indebted to environmental crusaders, and we should join them so as to ensure that poorly planned ventures which would endanger the environment are brought into line. Granted, environmental enthusiasts have sometimes mounted great prohibition campaigns based on too little evidence, or traded the dangers of pesticides for food scarcities. If such things go too far, the public becomes responsible and they pay accordingly, in energy brownouts, high food prices, or shortages of materials.

Wide environmental spoilage and pollution are made up of millions of small contributions from chimneys, vents, drains, sewers, runoffs, exhausts, radiation, and noise. Dangerous conditions are usually the result of unplanned crowding, congestion, and overpopulation in limited areas, and of irresponsible disposal of waste with no adequate treatment. We all contribute, and we all object, but our objections must be based on adequate information, otherwise the system will not be self-correcting. A high level of public information on such matters is called for, because our profit-oriented economy does not control environmental conditions without external insistence.

In the United States, and in many other countries as well, there has been plenty of pure air and water, and a new location to be sought if the old one is spoiled. As the population grows, this is no longer true. We must hope that local difficulties will always teach us this before any widespread environmental damage is done.

Someone has said, "Don't criticize the farmers with your mouth full." If beef prices are high, we should be aware that a pound of chicken costs less than one-half as much grain and land to produce. Our diet calls for a total equivalent grain consumption per capita over four times as large as that in India. Our demands for animal protein result in great quantities of animal waste, another pollution threat to our air and water. Decisions based on ignorance of such facts are likely to be faulty.

If we take insufficient corrective action, several environmental limits could be reached before the year 2000. We must keep in mind that another 50 million people are almost sure to be added to our U.S.

population before that date. Fortunately the warning signals are already here, and action programs are in evidence. We have an energy crisis, local but acute examples of pollution to be corrected, and much higher prices for food, electricity, and fuels. Worldwide scarcities and resulting inflation occur because supply cannot keep up with population and rising living standards, especially when we greatly increase nonsupplying activities, such as personal services and overkill armaments. We are heeding these warnings, but each individual must expect to change his or her habits and demands accordingly, otherwise the environmental limits *will* be reached in more and more American communities within our lifetimes. We need to apply the technologies we already know, to preserve our environment; and several new technologies are needed, that for nuclear waste disposal, for example. We are aware that combustion engines that do not seriously pollute are available; but instead of demanding nonpolluting engines, we add fuel-wasting and materials-expensive gadgets to the millions of polluting engines we continue to build every year, and then we set ahead the deadlines for requiring near-zero pollution. This is dangerous.

Efforts to ward off the more distant threats to our environment should certainly continue, especially in connection with the development of new energy sources and fuel conversions. If all processes and plants are developed with a view to environmental protection, no pollution emergencies will suddenly appear when large-scale operation is finally attained. Recovery of land after strip mining, disposal of ash, refuse, and shale residues and of noxious gases and vapors from any roasting process, storage or disposal of nuclear reactor residues, treatment of hydrogen sulfide from geothermal sources, and of sulfur oxides from combustion—these are a few of the pollution dangers that must be overcome *before* full-scale plants are placed in service. Economic support must be worked out where the profit system fails to provide it. The other alternative is a great scaling down of our energy-intensive pattern of industry and life.

There are many other possibilities for avoiding environmental limitations in the energy field. These include economic recovery of pollutants from stack and exhaust gases and from industrial effluents. The breeder reactor, the fusion energy plant, magnetic processes such as MHD, conversion units such as fuel cells, and even the large-scale production and use of hydrogen rather than hydrocarbons, have their environmental implications, as do solar power and wind and ocean sources. Present generations must be involved in these developments, even if their large-scale use can only be realized 25 to 50 years from now.

Environmental limits must be rigorously observed during the tremendous expansion in food production necessary for an added population of two or three billion on the earth before the year 2000. In particular, the contamination of water runoff by pesticides and fertilizers must be guarded against. Since high-yield grains demand more fertilizer on the same land, efforts should be made to provide these great quantities of fertilizer with minimum environmental disturbance and with less dependence on scarce raw materials such as natural gas. Results of research along these lines would be immediately usable.

A major objective in this book is to show the interdependence of our nation's current problems. The environmental field provides first-rate examples of these interrelationships and also emphasizes the fact that the solution of a largely technical problem in pollution control may have wide economic, political, and social implications. In devising antipollution regulations the sad fact is that so much time is spent in arguments about the acceptability of some of the control measures that the environment deteriorates seriously before correctives can be initiated. However, it is recognized that hasty action can result in the setting of impossible standards. It may even be necessary to experiment with temporary or interim legal measures and then to modify them as the economic and social consequences develop. Environmental limits are definite, and they arrive rather suddenly, as is demonstrated, for instance, by the air pollution in Los Angeles. Legal measures to deal with a given case of pollution involve so many economic and political problems, however, that a reasonable solution may take years, and most of the conditions change in the meantime.

A specific example is that of trying to control the air pollution caused by emissions from motor vehicle exhausts. This is by far the largest source of pollution in most city centers. It should be noted first that, while most cases of air pollution are local, and local ordinances and enforcements are required, the causes of air pollution are such that both state and national regulations are also necessary. Accordingly, in this case, Congress decided in the late 1960s to implement the commitment to clean up the air in cities by establishing technical specifications for maximum allowable contaminants from engine exhausts.

This policy of pollution control through limiting the dumping of wastes was already common, but no previous regulation had so involved every citizen, every business and industry, and every department of government. By the law that set legal limits for exhaust emissions of new 1968 cars, plus a program of extending this to 1975 and beyond, every vehicle owner was promptly alerted and became attentive to the means by which this was to be accomplished. Compliance would cost money, hence every present and future buyer of a new vehicle had an economic stake in the whole operation. Obviously, the simple specification of allowable percentages of carbon monoxide, hydrocarbons, and nitrogen oxides in vehicle exhausts posed a technical problem, mainly for the automotive manufacturer. But how can a federal emission reduction program which becomes progressively tighter over a period of a decade or more possibly be fair to *every* purchaser of a new vehicle and to manufacturers and dealers also? Does Congress, or anyone else, have the knowledge and wisdom required to specify the combustion within a vehicle engine so that accurately specified percentages of carbon monoxide, hydrocarbons, and nitrogen oxides will not be exceeded by any new car, truck, or bus of any size and model?

This law opened a Pandora's box, and more ills appeared every year. Enforcement problems proved to be awesome. Since start-and-go city driving is much worse than arterial or freeway driving, especially as regards carbon monoxide emissions, some arbitrary combination of driving modes had to be selected for the compliance tests. One-half each of urban and rural driving were assumed, but this was later changed to an "average urban driving cycle" on a dynamometer. Two other test cycles have been used, the "key mode" including idling, low cruise, and high cruise, and the "idle mode" in which all exhaust emission measurements are made with the engine idling. During one period the EPA test included acceleration and deceleration. (See discussion of emission factors in Zerbe and Croke's *Urban Transportation for the Environment,* Ballinger, 1975). Many cities have supplementary codes. Under such jurisdictions all tests are likely to be made with the engine idling, because this is the least expensive since no dynamometers are required. It is difficult enough to obtain representative samples of the exhaust gases and repeat readings of the instruments. The added variations inherent in different test modes leave added room for argument about compliance with a standard that changes from year to year and implies accuracy to at least two significant figures.

To make it worse, it soon became evident that this specification alone was insufficient to accomplish its intended purpose of reducing air pollution in heavy-traffic urban areas. Moreover, after the oil embargo in 1973 another purpose was added, and it was urgent, namely, fuel economy and conservation. After purchasing the vehicle the user might be faced with legal limitations as to where it could be driven at what speeds, where it could be parked, and the type of fuel to be used. Mixed local and federal regulations added to the difficulties.

For automotive manufacturers the long-term question was whether the increasing requirements for the next decade or more could be met by modifications of present engines with added external pollution-control devices. Might it be necessary to substitute an entirely new type of power plant instead of the old Otto-cycle engine, perhaps a gas turbine, a diesel cycle, a Stirling cycle, or a Rankine cycle? Possibly a stratified-charge engine would do, with fuel injection under computer control. The production line problems were frightening, and what would all these added costs do to the market? Could the oil companies meet fuel specifications, with the new fuels available everywhere?

The difficulties forced adoption of less stringent federal interim standards, and the final outcome of the program is yet uncertain. Occasions for argument and disagreement keep increasing. The changing cost relationships due to price increases have affected the entire enforcement program. Checking of vehicle performance for the user was included in several local pollution control programs, adding to user cost and inconvenience and bringing in all older vehicles as well as the newer ones. Even for simple tests the costs are high, and any method of payment other than a uniform fee falls heavily on the poor and the unemployed, since they drive the worst cars. Further cost complications came with the great increases in fuel costs, emphasis on conservation, and government insistence on a high level of vehicle maintenance. The objectives and the priorities became confused. How much of the cost is for fuel saving rather than for environmental improvement? In any case, the result must be that all vehicle owners continue to pay an increasing price for transportation of people and goods.

The cities with the greatest pollution problems were besieged by environmentalists and plagued by suggestions from every quarter. Since the most pollution occurs at start-up and at slow speeds, special traffic lanes should provide continuous traffic flow. Parking should be limited or prohibited in dense-traffic areas. Staggered working hours and night deliveries of goods should be required. Tolls should be charged on certain urban routes. Mass transit should be subsidized.

Vehicle changes were urged by many. These included much smaller cars, electric vehicles for short runs and deliveries, and fleet change-overs to bottled propane gas or hydrogen fuel. Most of the changes cost money, which the user must eventually pay. Smaller cars but more passengers, fewer trips, more frequent maintenance, were urged on the private car owner. The trucker must observe lower speed limits on the road and take designated routes in the city, both of which take more time.

The original standards for permissible emission of exhaust contaminants were based on a program finally to reduce carbon monoxide to 4% of a precontrolled average, hydrocarbons to 5%, and nitrogen oxides to 7%. Late in 1974 it was estimated that, in spite of relaxed interim standards, at least a dozen of the largest cities will still be in violation in

1977. If enforcement can finally be accomplished, a question remains as to whether the program will actually result in the improvements in urban air quality it was designed to produce. At the end of the program there must also be an evaluation that takes into account all those interim changes in external conditions that bear on the original assumptions. Then the question must again be faced: Are the target levels adequate? Whatever this final evaluation shows, many side effects are already evident, including the fact that some measures, the catalytic converter, for example, produce new contaminants which are objectionable.

The entire plan for reducing pollution from vehicles has been costly, and the price increases mean not only a higher percentage of the family budget goes to transportation, but local taxes have been increased to support local enforcement, larger traffic departments, and mass transportation. Automobile comfort and convenience have been reduced, and farmers' costs and food prices have increased. The higher costs of moving all goods have added to inflation, penalizing those who can least afford it. These cost increases and the life-style modifications they demand are of course not all chargeable to environmental measures. It is difficult to say how much is due to price gouging by the Middle East oil cartel, but the demands of pollution control did produce an upward pressure on prices of all vehicles and all fuels and thus have added to the cost of services for transportation, electric service, heating and air conditioning, and all energy-intensive industrial processes. Conclusions of the 1974 report of the Ford Foundation Energy Policy Project (Ballinger, 511 pp.) added a concern that the necessary U.S. energy policies have increased government intervention and government planning and that, even if the energy growth rate were to be markedly reduced, pollution would still be beyond the capacity of our environment.

Preservation of nature's best environment is a valuable service to everyone, but this example shows that such efforts are costly and produce many side effects. It even takes time to arrive at a reasonable and consistent policy, and much more time to accomplish the desired results. The programs increase the fraction of the working population employed by government, and the fraction of the industry payroll that must be classified as overhead. When manpower and capital are diverted to activities concerned with pollution control, costs of products are increased and prices adjusted accordingly. Every individual then pays, directly, indirectly, or both.

In this example the higher costs impinged on two of the most necessary services required by our industrial civilization, the transport of people and things and the energy for all uses. Even though the main objective was to reduce air pollution in cities, the farmer was hit hard.

Higher costs of vehicles and of fuel thus added to everybody's food bill and affected every family.

The entire environmental control field should be reviewed periodically to determine the extent to which the environmental limits are being approached. All federal activities in the field are reviewed yearly in the annual report of the CEQ, which is required by law to be submitted to the president in December of each year. The fifth annual report, "Environmental Quality 1974" (published in 1975), was a 600-page document summarizing environmental conditions and trends. It emphasized the prime relationship between environmental quality and population growth and cited the energy crisis as the major event affecting the pursuit of environmental quality, with the added complication of increased energy demands for expansion of agricultural production to provide food for a fast-growing world population. This report gave attention to two important questions, namely, the costs of environmental control programs, present and future, and the prospective rates of future growth.

This CEQ report tabulated the best available statistics on 1973 costs of pollution control and projected the cumulative costs through 1982. The total of 1973 costs, tabulated separately as capital costs and those for operation and maintenance, were given as $14.5 billion. These covered air pollution, water pollution, and solid waste controls only, as the corresponding costs for other types of pollution were not available. Approximately one-half of the total was for water pollution control.

Projected 10 years, the corresponding 1982 costs were estimated to total $45.6 billion. The *cumulative total* estimated for the 10 years was almost $320 billion, including annual capital charges of $132 billion, using amortization at 6% for public investments and 8% for private investments except automotive equipment, which was amortized at 10%. While these data are admittedly approximate, and costs of radiation and noise control and land reclamation are not included, they do give some indication that control of pollution must be paid for, and at an increasing rate. For comparison, the government publication *Construction Reports* gives the total value of *all* new construction put in place in 1973 by all privately owned public utilities in communication, transportation, and energy as $14.7 billion, which is almost the same as the total cost of pollution control the same year.

As to future growth this report includes the proposal of a slow growth plan for *total* U.S. energy consumption, projecting this growth at an average rate of 1.8% per year from 1970 to 2000. Previous studies have assumed higher rates of growth (see Chapter 3, Section F, and Tables 18 and 19). The curious basis for this proposal, called the half-and-half plan, has made it the most radical proposal to date. It justified the very low rate of energy increase on three expressed considerations: (1) "The production and consumption of energy is the single major source of environmental degradation"; (2) "the era of energy growth through exploitation of domestic supplies is over"; (3) "the U.S. must achieve a capacity for energy self-sufficiency consistent with national security." The half-and-half plan is based on what the report calls the "historic growth rate" of a 1.4% increase in energy consumption *per capita.* This is to be equaled in the future by a 0.7% increase in consumption per capita and a 0.7% saving through the elimination of waste and improvements in efficiency. Since the savings through conservation involve no increase in the fuel supplied, this "bootstrap" plan provides only a growth of 0.7% per year in the fuel supply per capita. If population growth averages 0.8% per year, the net increase in total fuel supplied will be 1.5% per year. (For comparisons see Tables 18 and 19.) This CEQ half-and-half plan was developed in March 1974 "to stimulate serious examination of the opportunities open to our nation through energy conservation." It was designed "as a serious long-term national program to conserve energy and meet the needs of a growing economy". One explanatory remark was: "Since the mid-1960's the per capita growth rate has exceeded 3% per year, a level which cannot be sustained".

This half-and-half plan provides for *very* slow growth indeed and further expects consumers to increase the efficiency of their utilization of energy by an average of 0.7% each year for 28 years. This would be easy at first, but would be nearly impossible in the later years.

In view of the mood of the American public and their representatives in Congress it could be said that many have not even accepted the existence of an ongoing energy crisis. Current reports on economic expectations indicate a very slight chance 'that a 0.7% increase in energy growth per capita will be accepted. It should be noted that the three members of the CEQ were conservationists and ecologists.

Accompanying the presentation of this half-and-half plan the estimated contributions from various energy sources in the year 2000 were given roughly as follows: nuclear 30%, coal 27.5%, petroleum 21%, natural gas 16.5%, hydro 3.5%, and solar and geothermal 2.5%. The percentage for coal included shale oil fuel. About 40% of the petroleum was assumed to be imported. The net energy consumption at the point of use was estimated at 85×10^{15} (as compared with 140×10^{15} in Table 18).

The three examples summarized above, that is, efforts to reduce air pollution from engine exhausts, the present and future costs of pollution, and the CEQ proposal for very low rates of growth in energy consumption, all indicate that there are several factors besides the limitations of the environment that govern programs for the control of environmental quality. In fact, the economic and political barriers and the need to educate the public will extend any environmental program over a period of several years, during which much of the action will be limited to dealing with emergencies. While general deterioration of the environment is no immediate threat, the necessities of providing for a growing population are immediate, and they will continue to pose new challenges for each generation.

E. THE POLITICAL LIMITATIONS

Political limitations are often the governing factors in efforts to deal promptly and adequately with economic difficulties. Democratic bodies such as the U.S. Congress and the UN General Assembly deliberate long and thoroughly to force the examination of all aspects of proposed actions, but they certainly lack the ability of the lone executive to respond quickly with a single-purpose program. The mere thought of setting up adequate measures to deal with crises like those in energy and food reminds us of past failures, and the delays caused by the seeking of advantages and the making of trades. "The best prophet of the future is the past," and the reaction of Congress to the energy program recommended to them by President Ford has been less than enthusiastic. Some partial steps have been taken, but action has been slow and inadequate.

In economic matters market forces and responses, such as demand, supply, and price, act rather promptly, and no long deliberations are involved. Too often legislative bodies are unduly influenced by political ideologies. They are also inclined to solutions that apply direct prohibitions which interfere with economic self-correction. Typical examples are price controls, quotas, embargoes, rationing, and subsidies. Such measures usually reduce the individual initiative allowed in economic action, and they are disastrous to orderly production. There are endless examples. Some that are pertinent here and which have produced unintended and undesirable effects are price controls on natural gas, rationing, prohibition, quota laws which resulted in extensive black market operations, and many of the welfare and unemployment subsidies that make idleness look better than the jobs available.

Arbitrary increases in money supply and debt and the fixing of interest rates make certain programs possible, including consumer purchases, military and public works expenditures, and military aid; but they may also induce opposite effects, such as the recent drastic slump in housing and private construction. There is no better example of the disastrous effects of price fixing and quotas than the 1974 oil crisis, a joint political effort to make up suddenly for generations of near-exploitation of several poor and backward countries by the industrial nations. The unsettling and adverse effects of trying to solve economic and trade problems by political fiat are well-known to most lawmakers, and this very knowledge results in serious delays and inaction. Sudden dislocations such as major crop failures with resulting starvation, or prices being tripled by the oil cartel, demand early action; yet action is delayed by political arguments. Worldwide efforts in the direction of population control, and those that would establish "laws of the oceans," have certainly been hampered by political considerations.

There is no intention to suggest here that the desirable political measures for dealing with the four related crises, overpopulation, environmental damage, energy supply, and food shortages, are simple and well known. But inaction is one great danger, and arbitrary interference with the flow of goods and with the processes of their production is another. Solutions are long-term and call for wide cooperation, and since the ills themselves are largely caused by one of the drives that preserves the individual and the race, namely, self-protection and selfishness, it is necessary to convince thinking individuals that certain of their actions must be modified in the interests of their fellow humans. What and how much to eat, procreation, what to do with wastes, how to protect from heat and cold, when and how to travel, how to share nature's bounties— these are the kinds of basic individual decisions involved. Some of the means toward the solution of the four crises involve limitations on personal luxury, avoidance of waste, limitations on the tools of conflict and war, and sharing with the less fortunate.

How can the desired ends be attained? Again we must start from past experience and present facts, and these are so extensive and so complex that they call for a major reeducation on our part. This reeducation to modify our style of life and philosophy of growth will take time. The solutions will come *after* the educational process and after the political process of decision and action. In the meantime, political limitations will become more conspicuous, and political wrangling, inaction, and mistakes will hamper effective results.

A large share of all political maneuvering, both in business and in government, is motivated by selfish purposes; but partisan objectives often make an even greater contribution to unwise decisions and actions. Partisan pressure might be called "group selfishness," but this is an oversimplification, and underlying forces and motives should be recognized. In the business sphere, profit and competition are driving forces; and in government the securing of advantages for one party, for a local area, or for a department or agency may be paramount. Continuity of office or influence is another strong motivation in both private and government sectors. Partisanship is well recognized and easy to find. Everyday examples include the activities of congressional committee chairpeople who quietly secure more and more government facilities for their city or state, the jurisdictional selfishness of an administrative department, the president tilting with a Congress dominated by the opposite party. Another deplorable example of partisanship is the political manipulation by congressional, staff, or committee members of a project concerning which they have no background in the economics, history, or technology of the subject at hand. Even consumerism is partisan. One sad picture today is the political advantage taken by the consumer advocate, and the resulting likelihood that Congress may enact political controls that may make it impossible for whole industries to prepare adequately for future demands, thus forcing them into a position similar to that of the railroads today. The energy utilities are a current example.[58]

The fact that our political structures, both national and state, are organized geographically is inclined to result in a project or a proposal being approached from a geographical viewpoint rather than a functional one. Most of our public officials have either a legal or a business background. They are accustomed to recognizing the influences of locality, but not the requirements imposed by technology or rigorous economics. This deficiency in information within the government should be corrected.

When a corporation management is seeking competitive or profit advantage, or even when it is striving for recognition as a public servant, it is more than likely that it commands competent and skilled assistance in economic and technological matters. The efforts of the large oil and gas companies, as indicated by their annual reports and their national advertising, is a case in point. Such specialized assistance in decision making is less available to Congress, or even to the president, although there have been some recent improvements in this regard. The activities of the National Academies of Science and Engineering, the NSF, the BEA, and the environmental agencies include such consultation.

Long-range improvement is needed in the contribution of economics, technology, and economic history toward reducing the political obstacles to wise decision making. Effective action in the fields of energy,

[58] See "For the Utilities It's a Fight for Survival," *Fortune,* **XCI** (3) 97 (March 1975).

food, development aid, and environment is dependent on the applicable information and knowledge possessed by legislators, executives, and managers. Our educational system has an obligation here. The complexity of our industrial society demands incoming managerial talent which has studied science, economics, and their practical applications. College curriculum planners and student advisers can accomplish much and should be persuaded to so direct and convince their students.

Political limitations due to inaction followed by arbitrary decisions, or due to individual or group selfishness, can be minimized by zealous leadership, with occasional necessary changes in organization or procedures. Overbearing committee chairpeople and group filibusters can suddenly succumb to such changes. But to accomplish an alteration in life-style or philosophy is a long-range educational project. So also is the diffusion of accurate information on such involved subjects as energy production, food supply and distribution, and the growth of populations and their demands.

Chapter Eight

Adjusting to
Future Demands

A. *CHANGES IN AMERICAN LIFE-STYLE AND EDUCATION*

1. *The Pressures for Change*

Style of life is a matter so personal and individual that it is difficult to take a broad viewpoint. Life-styles vary greatly among countries, even among different groups within a single community, and each may prefer his or her own.

The younger generation is critical of the life-style of their parents, and they strive to establish something different, presumably better. It is admitted that life preferences and habits do depend on age, vocation and station, and personal differences. Freedom and comfort are highly prized by almost everyone, however. There are other common elements, but the point to be made here is that these common elements have been changing rapidly and are due for some more changes.

In the industrial nations in particular it is gradually being recognized that the several interrelated trends emphasized in this book, with respect to energy, food, environment, population density, and growth, will make irresistible future demands on our personal style of life. These represent controlling factors in everyone's future, but adverse effects can be largely avoided by anticipation and planning based on full information.

An Arab oil boycott, a famine in India, a crop failure in Russia, the appropriation of U.S. corporation properties in Cuba, Chile, and Venezuela—all these incite resentment when they result in high prices for gasoline, bread, meat, sugar, or copper. Air pollution, bumper-to-bumper traffic, crowded recreation areas, and racial clashes, are regarded as intrusions on our good life which have been imposed by someone else. Several future threats have arisen because of our *inattention*. They may be the result of the shift from rural to urban living, the change from goods production to service occupations, or the departure from provincial living in a single location to a mobility that approaches that of a nomad.

Some of the broad changes have resulted in habits of luxury and waste. We have enjoyed big cars for 70-mile speeds on smooth thruways and then have used them for countless trips to the corner store. We have owned power boats, snowmobiles, and sports cars; but high prices of gasoline were not ours to worry about. Former luxuries of kings and captains of industry have now become necessities for the great American middle class. We begin to realize that our reckless use of oil, natural gas, and electricity, our throwaway gadgets, toys, and packaging, our steak dinners, and our miracle fabrics have been using up so much of the world's materials that we are depriving others and threatening our natural resources and environment. Where else are fully equipped summer homes and 5000-mile holiday trips so common?

Many of the pressures for change in our habits are arising because the rest of the world is learning about the contrasts between our material standards and theirs. They now have radio, movies, and television which picture our greatest luxuries. The common cries of "imperialism" have resulted largely because the individual market economies draw much more than their share of the world's valuables from everywhere. No matter that these raw material industries have been developed by Western initiative, the supplying countries want a larger share, which means some changes in what we get and what we pay. Other cartels promise to follow the OPEC.

By examining twentieth-century American development and the origins of our vast economic advantages, perhaps we can recognize that certain accelerating trends must be modified if we are to ward off the dire consequences described in the Club of Rome reports (see next section).

At least one major source of our advanced position in material prosperity has been our outstanding success in developing mass production, first in factories, then in agriculture, and even in education and communication. Specialized mass production demanded specialized planners, organizers, and operators; and our educational system adapted quickly and eagerly to the role of training more and more specialists. Starting more than 100 years ago with increased attention to vocational education (the Morrill Act, 1862), universities gradually established large specialized schools for agriculture, engineering, management, and business. The mass-production techniques of the telephone, automotive, and appliance industries were studied and applied, even in the universities themselves. Practitioners finally were so specialized that they could communicate only with their own kind.

Early in the 1950s some educational leaders, after observing what had happened in World War II, began to advocate broader education, even for the specialist. Promoting greater attention to fundamentals and less to the techniques of quantity growth, educators soon became aware of support from an unexpected quarter. The younger generation in the 1960s not only rejected some of our materialistic goals, they also saw little advantage in our ambitious, unlimited growth.

The need for increased emphasis on quality and objectives instead of on quantity and specialized techniques was accented by a sudden interest in man's natural environment. While this movement may have been started by nature lovers and crowd haters, it set loose a flood of printed words and propaganda which generated great support for legal means to protect the environment.

This mixed revolt has weakened the influence of the business establishment. Reappraisal continues, and the middle road finally begins to appear. Economic pressure has diverted both the specialist and the dreamer to a more versatile program. The new trends reverse some of the most highly regarded objectives of the first half of this century. The new purposes are less specialization, broader education, less attention to quantity output and growth, and more emphasis on quality. The aims include more individual participation and less passive, "spectator" absorption, and less conquering of nature and more cooperation with her.

2. New Education and Economics

Since both formal education and self-education evolve only gradually to meet new demands, any expectation of a sudden response is likely to be disappointed. One reason for this is the persistence of the habits of thought and convictions acquired in the early years at the primary and secondary school levels. While the local and global difficulties outlined in previous chapters have been called crises, they are really long-range problems that will respond only to educational efforts. They require new points of view within large segments of the population.

Many educators consider that planning for the future is an emphasis too practical and too mundane to deserve much attention in their curricula. In fact, some believe that economics, technology, and even the sciences are specialties which should be omitted from outlines for a general education. There is now some indication that this rather smug attitude is being changed by the evident urgency of such problems as the control of population growth and the wise use of the earth's resources. But it is not easy to alter the tenet or promise of unlimited growth.

In America especially the conviction that unlimited growth is good, and desirable in almost every instance, is inbred. It is taught from preschool to graduate school, and by the written, spoken, and electronic word. It is praised in every review of past performance and every prediction for the future. Enthusiasm for growth is contagious, and to counter with the idea that slower overall growth is necessary and just as desirable is a tremendous educational assignment.

Fortunately, a good start has been made, but the response is largely limited to emergency measures. The public is still skeptical about the energy crisis, and the need for slowing the growth of population is also accepted very reluctantly in many quarters, even in the face of famine. Our established process of "government by emergency measures" keeps everyone so occupied that minimum attention is paid to long-term planning.

Admittedly, such questions as fuel rationing, reduction of unemployment, and dealing with runaway inflation or with high prices and local shortages of food must have first attention from those personally involved. Planning for the distant future is likely to be neglected by a young family or even by a business because 10 years or more seems like too long a time to plan ahead. Questions involving the environment, the future energy supply, and food for a fast-growing population thus call for an approach that starts with a broad education on the needs and the alternatives. It must be appreciated that it takes several years to build new plants or to transform desert areas into farms.

We now face a convenient interim period, the balance of the twentieth century; and if wise plans are made and carried out in this interim, we can hope that future generations will thank us. This emphasis on the years before 2000 is already well started. In education and economics attention is rapidly shifting from emphasis on techniques and the tests of short-term accomplishment to greater appreciation of basic principles and more concern about the environment we are creating for the next generation. Devotion to perfecting the techniques of production, maintenance, and growth is giving way to careful appraisal of, and responsibility for, the ways in which the techniques are being used. The human value of one's results becomes a concern of the educated technician, engineer, or economic analyst and planner.

As specialization becomes less of an end in itself and quality of life receives more attention, the quality of things and of services, and of the jobs that provide them, receive more consideration. So also does conservation of the materials from which things are made, and of the pleasant and healthful environments nature has provided.

Present shifts to the new emphasis are encouraging, both in fields of education and in the setting of economic objectives. Futurist societies and publications have appeared in many countries. They are concerned with the human environment and with the effects of rapid population growth. When the energy crisis suddenly became acute, the study of the future of energy supplies was very quickly embraced. Even in the universities, the study of environment, ecology, and energy are being included in the fields of economics, engineering, and management. Large corporations are appointing long-term planners who are freed from the concerns of day-to-day production, marketing, or finance.

Broader education and reeducation for adults has been forced in part by rapid technical progress, but these are now being widely supported for humanistic values. Environmentalism and conservation are endorsed for similar reasons, and because of the growing conviction of their necessity. A simpler life is being cautiously advocated without reference to austerity. Many members of the younger generation are leaders in these desirable changes in educational emphasis, and the economic recession has endorsed the necessity for change.

While a start has been made, the objectives will not be attained unless action is accelerated. More business people and educators must be convinced of the need for early adjustment to the new demands. It is urgent that full information and the supporting argument be spread much more widely.

B. THE RISKS OF HUMAN DISORGANIZATION

1. Short-Range Risks

Disorganization is one of the most potent enemies of a complex structure like the economy of today. Fortunately the near-future dangers are small, and the opportunities and challenges within the system very large.

While the risks are not great and the ills are distant, each should be fully recognized. One danger is the distortion of trade for political ends, as has been so devastatingly shown by the oil supply example. When the primary cost of producing Middle East oil is only 20 cents per barrel and the selling price is boosted to over $10, economic relationships are hidden by political opportunism. In 1974 the net profits of the OPEC were $60 billion, and they are still growing. This is a serious disorganizing force that the world economic and financial systems cannot long withstand. With the entire world relying on petroleum energy for domestic, commercial, industrial, and agricultural uses, not to mention the military, both the everyday operation and the future plans of every country have been suddenly thrown into confusion. Gasoline rose to a range of up to $2.50 per gallon in several European countries, making its cost prohibitive for many normal uses. The same was true for fuel oil for heating and for industrial furnaces and ovens. Confusion and disorganization are a real threat.

The United States has been guilty of another example of mixing political pressure with trade. This was the law that made most-favored-nation status for Russia dependent on her Jewish emigration policy. While neither Russia nor the United States much needed the trade, and other nations were not harmed, the same principle is involved.

Another risk in today's closely integrated world is the magnification of local disturbances, as in the Cyprus affair. A disagreement and struggle that should directly involve no more than the equivalent population of a single city or county elsewhere, becomes a world concern, distorting economies and even threatening confrontation and war between world power blocs. It is still not clear why treating the wars in Southeast Asia as local disturbances would not have been a wiser course.

A third disorganizing risk to orderly economic operation is the pattern of terrorism so well demonstrated by the Palestinians and the Irish. Violent reaction to social or economic injustice is certainly not new, but the weapons and the havoc they can create are a widening threat to our international and internal relationships. Even in the United States a good example is provided. A few instances of airplane hijackings made necessary the search and inspection of every airline traveler and all his or her personal effects.

2. Long-Range Risks

The long-range risks of disorganization can be visualized from either of two extreme or opposite viewpoints, overgrowth or stunted growth and stagnation. Any widespread movement of noncooperation can result in economic stagnation. Minor local examples are strikes and lockouts, but large organized opposition to ongoing systems tends to act in the same way. For example, environmentalists opposing the location of a power plant or an oil refinery in their own neighborhood, and consumers organizing to enforce uneconomical pricing of a utility service, both eventually lead to failure of the essential supplies, with the attendant disorganization of the economic community.

Refusal to admit the existence of shortages, of energy, metals, and food, threaten to produce stagnation and disorganization when the shortages become acute. Uncontrolled growth is rightly opposed, but the slogan "zero growth" has been an unfortunate one because it offends one of the great faiths of mankind and appears to be in conflict with both nature and technology. This connotation breeds great opposition, which in turn disorganizes the efforts of those who are trying to modify the dangerous trends of today. The promoters of zero growth should emphasize that what they mean is growth to equilibrium, taking the environment into account, or else organic growth to the equilibrium of maturity, as opposed to unlimited growth, which in nature is pathological.

Overgrowth, the second objectionable kind of long-range growth, can be vividly emphasized by the simple extrapolation of present growth trends into the future for 100 years, or even less. Computer models for such an exercise have become common. Perhaps the best known are the two reports to the Club of Rome, which have already been mentioned (see footnote 3). The 1972 Meadows report, *The Limits to Growth*, was translated into more than twenty-five languages with circulation in the millions. The 1974 Mesarovic-Pestel sequel, *Mankind at the Turning Point*, received an even more impressive reception.

The world always listens to predictions about the future, and *Limits to Growth* soon achieved the distinction of being actively promoted in a dozen countries, and severely criticized in just as many. Following widespread international review by the press, radio, and television, the book was discussed at several conferences, awarded recognition by prizes, and soon became the focus from which emanated several other studies of the long-term future. As compared with the many national prediction studies made in the United States and elsewhere, the global approach of the Club of Rome appealed to world-minded thinkers in such groups as the UN and the OECD; and more research from a similar viewpoint was soon initiated. Such organizations as the International Federation of Institutes of Advanced Study and the International Institute of Applied Systems Analysis are now engaged in world problem studies of similar import. The latter institute held a seminar in 1974 at which global economic models were thoroughly examined in about thirty reports.

This first Club of Rome report starts by relating the decision of that body to undertake a project on the "predicament of mankind" and the problems troubling all nations, such as "poverty in the midst of plenty, degradation of the environment, loss of faith in institutions, uncontrolled urban spread, alienation of youth, rejection of traditional values, inflation and other monetary and economic disruptions."

The team proceeded to examine five factors: population, agricultural production, natural resources, industrial production, and pollution. These were selected as the factors that ultimately limit growth on this planet. In setting up the model, it was emphasized that the variables are "dynamically interacting elements." The 200-page presentation first summarized world economics and resources. Then computed results were presented in graphical form, usually plotted against time, from 1900 to 2100. Final conclusions were that present trends, if continued, will most probably lead to "rather sudden and uncontrollable decline." They stated that these limits would be reached within the next 100 years.

The Meadows team admitted that such a conclusion is not new and that it is quite possible to alter the dangerous growth trends, but the sooner the better. They did affirm that basic changes in values and goals would be necessary if enduring "equilibrium by planned measures" was to be attained. Their conclusions sounded an optimistic note, however, and they looked forward to further studies.

The second report to the Club of Rome, *Mankind at the Turning Point*, is prefaced by recognition of a major threat to the human future, namely, the precarious stability of human existence in the face of continuous escalation of nuclear armaments. With the fond hope that this stability can be continued, the report then devotes its attention mainly to four current and unprecedented crises: runaway population growth, the energy problem, food shortages, and pollution of the environment. Methods for solving any one of these crises separately are recognized as inadequate, and even as an aggravation of the other problems. The real solutions are declared to be interdependent. It is naive to take a stand either for or against growth as such, and a plea is made for the abandonment of ideals calling for exponential growth. Organic-type growth is to be favored, and the current crises can become catalysts for change and thus blessings in disguise.

Against this background, the nature of these global crises is examined, and their ultimate results in an interdependent world are estimated by computation. In place of the usual arbitrary bureaucratic decisions, a holistic or systems approach is proposed, "which looks at the totality of all aspects of the problem." A computer model of the world system is then described, linking the individual to the ecology through a chain of social, political, economic, and technical disciplines.

While all actions are viewed in a global context, regional differences are taken into account by the authors, Mesarovic and Pestel, by constructing a multilateral model through the use of ten area models. Thus is produced a much more realistic picture than results from former one-world analyses. The stated objective was to depart from the single computer direct solution and to use the computer rather as a planning and decision-aiding tool. An immense number of calculations was used in assessing alternative future developments.

Several questions or problems were proposed and viewed separately. These involved the troubles of the developing nations, the results of population growth, the competition for scarce resources, especially oil, and the food supply problem. A large number of specific conclusions followed from the results of the various "scenarios," or alternative paths of action, in dealing with these problems. The computer results were shown in graphical representation in each case. A detailed examination of these results is strongly recommended to the reader.

The general conclusions are those that might be anticipated from a commonsense analysis, but they are given importance and verification by the highly detailed computer method used. The analyses were carried to the year 2025 or beyond, with results indicating that these current crises are not temporary problems but will continue, and that solutions can be developed only by attacking them in a global context through cooperation rather than confrontation. Traditional single-purpose attacks may be misleading, and certainly inadequate, for example, efforts to solve the food supply problem without regard to energy shortage and population growth.

One outstanding emphasis in this second report was on the urgency of early decisive action. Each "scenario" verified that delay will exact great penalties in human suffering and lives. Certain lessons as to the necessary revision of individual attitudes were pointed out, amounting mainly to a requirement for more unselfish values and considering benefits to future generations as well as to ourselves. This condensed report is also recommended for its "briefs" or status summaries on such subjects as world resources, agricultural yield and food production, fossil fuel reserves, and the growth and age structure of world population. The data are given in very compact form; and these sections, like the rest of the report, were prepared with authoritative collaboration and consultation.

These two reports to the Club of Rome are recognized as monumental contributions which will accelerate an awakening by pointing out the worldwide disorganization that would result from present trends. Their emphasis is on the urgent need for changing these trends.

This book you are now reading has the same objective of spreading that awareness, but with the purpose of focusing primarily on constructive programs within our own country. Prospects for the next 20 years are examined, based especially on what we have experienced and learned in the last 20 years. This book was completely outlined and well under way before either of the Club of Rome reports was published, and the objectives throughout its preparation have been consistent, namely, to provide a simple but fact-filled story of how we should be able to build on our past experience to overcome the threatened disorganization that would result from the continuance of present trends. After only a brief indication of future dangers, the purpose is to show just where the United States now stands in these crisis areas, namely, energy, pollution, food, population growth, and foreign aid, and to show how these areas interact and are closely related to growth. Any solid convictions about where we stand, and proposals for future programs, both demand accurate quantitative knowledge of the immediate past, hence the thirty-nine tables herewith which have been compiled to furnish just those figures that are necessary for adequate comparisons and planning for the future. It is agreed that early action is most important, but the immediate need is to provide the facts in each field and to suggest directions for change.

In these crisis areas, simultaneous efforts are gradually becoming nationwide and international as the information is disseminated that makes it possible for more and more people to understand the needs, the practicable remedies, and the risks and consequences of inaction. A mammoth educational effort is still necessary to ensure a wide understanding that will support the necessary endeavors to avert these long-term threats of disorganization. This book is intended as a contribution to that educational effort.

3. *Indicators of Disorganization*

There are three common indicators that measure departure from well-organized economic activity directed toward the good life. These are unemployment, inflation, and war. Each of these signals of disorganization can be roughly shown on a quantitative scale, but an adequate description of their adverse effects would call for book-length analyses.

Inflation in the United States in dollar terms is shown in Table 29. Such dollar values are computed in arbitrary ways and are subject to interpretation, as indicated in the short discussion in Chapter 6, Section D. Monetary inflation makes it possible to repay more cheaply the debts contracted by past excesses. Large segments of the population can protect themselves very nicely from major loss by inflation. There are automatic increases in income, inventory profits, and capital gains. The groups that suffer the most, however, the poor and the old, are ill-equipped to resist inflation by initiating changes in the system. In any case, inflation has a serious disorganizing effect.

Unemployment is a basic indicator, since it is evident that enforced idleness is both nonproductive and disruptive. Government statistics and methods of counting are admittedly arbitrary and even somewhat variable, and such national and yearly data also fail to indicate the acute local and factional disorganizations that result. Extensive inflation causes the unemployed to undergo much personal and economic suffering Minority groups and women are very evidently at a disadvantage in obtaining and holding good jobs, and some states fare much worse than others. The failure of our system to induct youth into the work force properly is indicated. Causes and effects of unemployment are beyond the discussion here, and the reader is referred to the very extensive literature on the subject, including the Annual Manpower Report of the President. It is notable that the United States has had more unemployment than several other developed countries.

Present and future wars absorb much productive effort, and this is not only a disorganizing force in the long-term economic struggle but constitutes a real tragedy for all humanity. Military planning and preparation, plus actual hostilities, divert vast amounts of available manpower and capital away from the struggle toward a good and peaceful life.

In all efforts to counteract the human losses traceable to unemployment, inflation, and war, it is unfortunate that the concentration is almost entirely on short-range remedies. Political leaders and legislators can devote only minor attention to constructive planning for the future because they must react to the clamor about emergencies this year and next. In a democracy they are reelected on the basis of their records and promises to solve today's problems. Top business executives must concentrate on dividends to stockholders this year and next, and on the accumulation of capital funds to protect long-term profitability.

The misfortunes of the United States in the mid-1970s can be attributed largely to loss of confidence. The economy suffered disruptions that resulted in a major reduction in consumer goods and services. The number of consumers continued to increase, but for many of them the good things in life were sharply curtailed, partly because of inflation, unemployment, and continuing military-mindedness.

C. THE PERSONAL OPPORTUNITIES

Whatever our concerns over slow future growth and shortages of energy, food, or living space, we must admit there never was a time when the long-term future offered more challenges to the individual American. There are so many things to be done, with plans already suggested; and there never was a better economic system or political climate inviting their accomplishment. Perhaps the limitations of capital and manpower, the risks of human frailty, and disagreement will slow each process; but there are opportunities for everyone, and freedom to choose the function he or she may prefer. The main prerequisite is to learn what is needed, and then to participate accordingly. There is even a challenging timetable, namely, to change the trends before the turn of the century. Dire predictions of disaster then give way to constructive enthusiasm which builds from existing foundations.

Is this viewpoint too optimistic, and blind to the great obstacles? Such a question contradicts our history and disregards the momentum already attained. It also ignores the demonstrated fact that when our country has a job to be done and goals to be attained, as was the case for the space program, leadership and enthusiastic support and cooperation promptly appear. Our complex system has attained its enviable position because of millions of competent participants, each with valuable ideas to contribute.

The vocational choices for youth today are endless within these programs for conserving and increasing our energy and food supplies, for preserving the natural environment, and for preserving the heritage of the American economic and political systems. In spite of slow future growth of the economy and of the population, opportunities for phenomenal individual growth and accomplishment are certain.

The initial impetus in a desirable direction was furnished by those forward-looking students of human affairs who, in the 1950s and earlier, published their well-supported warnings about the results of unrestricted growth of populations and of consumption. Even before that, conservationists, and later environmentalists and ecologists, reminded us with emphasis that the gifts of nature are limited. Such concerns have been voiced in many nations, and the UN has contributed substantial guidance through statistical studies and suggested programs. For decades, especially in the United States, geologists and engineers have been warning us that petroleum and gas reserves are limited. It has now been some time since demand for these products has caught up with U.S. production, so we are more and more dependent on imports. Federal government departments, the President, and Congress have studied the needs and publicized data and recommendations. The big foundations have explored the problems and made reports.[59]

For individuals, especially in the United States, we have arrived at the unique and fortunate position of having the future requisites well described, the threatening alternatives indicated, and the desirable programs outlined.

---

[59] Thought-provoking discussions will be found in Weintraub, Schwartz, and Aronson, Ed. *The Economic Growth Controversy*, International Arts and Sciences Press, 1973.

Compared with our grandfathers, who abandoned the certainties of home and friends to develop a continent, the great opportunities have instead been brought to us where we are. Even so, we have wide choices. Our great country itself offers change and variety, if that is what we want. In many small countries, moving to the city means moving to *one* city, because the country has only one big city. Change of climate is impossible; there is only one climate. Vocational choice is limited, because there is only one major industry. In our country the variety is endless. Each of the described crises, pollution, shortages, foreign aid demands, has a thousand facets calling for a range of skills involving different teams and different locations, yet our existing industries will remain and the opportunities they offer will continue. Our systems of private enterprise and democratic government are committed to individual choice and private initiative.

The scientist and the technologist have countless opportunities, especially in the fields of energy and resources, and in the new food technologies. Business people and organizers are needed for the new accomplishments. Lawyers must clear the legal difficulties. New courses are needed in more colleges. High schools can point out new directions for the impatient teenager. Many trainees will be needed in business and industry.

If more action is needed, and the way is clear, what is holding it up? Why is it so difficult for the enthusiastic individual to find his or her own place? How long must one wait?

There are several answers. No small part of the delay has been caused by the recession and the lack of confidence it represents. The democratic process we prize so highly is painfully slow. A prime example of this has been the reactions of Congress to the president's energy program. Both President Nixon and President Ford made great efforts to get a program under way; but it took 5 years of argument, and it is still moving slowly. Opposition and disagreement are always easy to arouse; but the paying jobs are on the constructive side, and they are being delayed. Nevertheless, it is good to take time to examine all sides of a proposal and to spread information widely to obtain the advantage of many viewpoints. In spite of delays in all-out programs many separate projects are under way, especially in the fields of energy, foods, and conservation.

Energy developments must and will accelerate very fast in the years before 1980. Increasing the efficiency of fuel usage will save billions of Btu and kilowatthours. Millions of people must be convinced of this, millions of fuel-using devices improved, and heat losses prevented. Better gas mileage for automobiles, reduced energy losses in the home, office, and factory must be accomplished. Thousands are already working at this, and millions more are needed. Better energy control for better economy offers long-term repayment anywhere fuel is used, but it is hard to get started and an army of people must cooperate to make it effective.

Every purchaser of new energy equipment should look beyond the temptation of cheap first cost. In the past we have bought cheaper cars, furnaces, industrial ovens, and energy process controls. Yearly fuel savings were not a factor because fuel was cheap. This must be changed, because fuel will never be cheap again. Thousands of well-informed people should be selling this idea, and thousands of others will design and manufacture more efficient units and then see that they are properly installed and used. This is no small assignment, even for American industry and purchasers.

Conservation of materials calls for a change in our habits of waste production. In addition to the reclaiming of valuable metals from waste, combustibles should be burned for useful heat. Goods and packaging must both be designed for minimum use of scarce materials such as metals and petroleum-based plastics. Conservation means new approaches and new jobs.

Keeping down the costs of food and improving the U.S. position as a world food supplier naturally involves first the farmer and the food processors, plus the government with its controls and regulations; but there are many other opportunities for the individual. Innovative ideas for improving the food supply have prompted a great deal of small-scale experimentation and production including emphasis on less expensive high-protein foods, crops of plants without soil, ocean foods and tank farms, fish and lobster hatching, and so on. The number of such small enterprises is much greater than is generally realized.

The immense size of the capital investment program required to carry out all the changes being considered here has already been discussed (Chapter 7, Section B). The next 20 years must see a great and successful effort in improving and expanding producing plants, their equipment, and their structures. Not only must these plants turn out synthetic fuels, more efficient energy equipment, and a great output of foods, but they must also supplant older capital facilities which were not improved or expanded during the recession.

The entire manufacturing industry, making cars, appliances, tools, equipment for construction, transportation, and communication, foods, and everything else, spent less than $40 billion per year during the 1968–1974 good times. Future amounts, for the energy field alone, are expected to be of the order of $500 billion in the next 10 years. This indicates that prerecession capital spending should be tripled (in constant dollars). Ten million direct jobs are involved in such a program. As a comparison, all nonresidential construction employed only about 2 million in the best of times; and the entire labor force was about 80 million. We must worry about a shortage of workers, not a shortage of jobs.

In the first few years of the new programs, by far the greatest number of people will be occupied in education, training, information, and personnel fields. The entire country must be sold on the needs and challenges. Professional, technical, and business training must be stepped up. Induction of workers and on-the-job training will occupy many. Schools, colleges, and mass communication media, and publishers will meet the demands; but it will tax their capacities.

The needs for foreign aid and for food exports will be expanding and calling for personnel. The government, voluntary agencies, and private business are all involved here. In agriculture there must be large-scale developments of farming on marginal land, clearing of land, development of irrigation, more fertilizer plants, and introduction of new fast-growing and pest-resistant varieties of grains, vegetables, fiber plants, and trees.

Much attention is already being paid to the new programs by the president, Congress, federal government departments, and state agencies. Starting with the energy program, government participation will be large, with a corresponding demand for competent personnel.

The more new programs are examined, the more it becomes evident that with a population growth of only 1% per year (or less), the statement that there are plenty of opportunities for everyone is not an exaggeration. Personal challenges are everywhere.

Projections into the future, such as those of the Club of Rome, support the contention that zero growth, both of population and of the economy, must ultimately be attained, since the resources of the planet are finite. Practical studies of how and when this can be accomplished, however, lead to the happy conclusion that there is still room to weigh alternatives and decide accordingly. The situation threatens no early calamity. Instead, it furnishes every individual a wide personal choice. Our selections are subject only to a recognition that unlimited growth is not the ideal.

There are opportunities for immediate personal contributions to the slowing of growth and the alleviation of its undesirable consequences. Every citizen has continuing choices with respect to the use of resources, the avoidance of pollution and, what is of the most basic importance, the rate of population increase. Every individual is challenged to learn what are the moves that will conserve energy, mineral resources, and nature's environmental bounties. The great personal advantage is that there are thousands of choices, so that each individual may consider alternatives and then freely choose in accordance with his or her own preferences and the needs of the future as well as the present.

Americans have a great advantage with respect to free personal choice, not only because of their rights and privileges as U.S. citizens, but because of the great natural resources of our country. These have been developed through personal initiative and advanced technology. We are determined to pass these advantages on to future generations, but this can be accomplished only if each individual pays sufficient attention so that he or she recognizes what the desirable moves are.

Appendix

Statistical Tables

Table 1. National Populations and Their Living Standards, Year 1970*

Countries of over One Million Population

Country	Population (millions)	Approximate population growth (% per year)	Life expectancy at birth (years)	Food (calories per day and % from animal sources)	Education (teachers per 1000 population)	GNP per capita (1970 dollars)	Electric power (kWh per capita per year)	Imports plus exports per capita (1970 dollars)
Afghanistan	17.1	2.4	38	2060, 7	0.4	80	20	17
Albania	2.0	2.9	66	2370, 14	1.2	400	450	—
Algeria	14.0	3.5	51	1890, 6	0.8	290	140	160
Angola	5.7	2.1	34	1910, 7	0.6	170	125	140
Argentina	23.4	1.5	67	2810, 34	6.6	1070	1000	140
Australia	12.6	1.9	70	3160, 45	4.8	2900	4300	730
Austria	7.4	0.5	70	3260, 41	6.0	1930	3800	865
Belgium	9.7	0.6	70	3220, 41	—	2540	3100	2300
Bolivia	4.9	2.6	50	1760, 14	1.7	200	160	7
Brazil	93.2	2.9	61	2620, 15	3.8	390	490	50
Bulgaria	8.5	0.7	71	3070, 13	4.0	800	2300	450
Burma	27.6	2.3	43	2010, 6	1.0	75	20	5
Burundi	3.5	2.3	37	2110, 3	0.3	65	6	13
Cameroon	5.8	2.0	41	2230, 5	0.6	180	200	82
Canada	21.2	1.8	72	3200, 46	14.0	3700	9500	1500
Central African Republic	1.6	2.2	35	2170, 7	0.3	120	30	31
Ceylon (see Sri Lanka)								
Chad	3.7	2.4	32	2240, 8	0.1	70	10	20
Chile	8.8	1.9	63	2530, 20	4.3	680	850	260

Country									
China (Communist, including Tibet)	780.0	1.8	50	2050,	9	—	90	—	6
China, Republic of (see Taiwan)									
Columbia	21.0	3.2	45	2140,	20	4.0	400	400	75
Costa Rica	1.7	3.7	63	2320,	15	2.4	500	600	130
Cuba	8.4	2.2	67	2500,	16	3.3	350	550	330
Czechoslovakia	14.5	0.5	71	3030,	27	3.1	1500	3100	530
Dahomey	2.7	2.6	39	2170,	4	0.2	80	12	28
Denmark	4.9	0.7	76	3250,	46	—	3100	3600	1580
Dominican Republic	4.0	3.0	58	2080,	14	1.0	360	225	116
Ecuador	6.0	3.4	52	1980,	18	—	270	150	76
Egypt	33.0	2.5	53	2900,	7	2.0	210	230	47
El Salvador	3.4	3.4	59	1850,	11	1.2	290	200	129
Ethiopia	25.0	1.9	39	1950,	9	0.2	70	20	12
Finland	4.7	0.3	73	2990,	45	7.0	2200	4500	1050
France	50.8	0.9	76	3210,	40	6.0	2800	2770	730
East Germany	17.1	0.0	72	3040,	37	8.0	1450	3900	490
West Germany	60.7	0.7	70	3150,	42	5.0	3000	4000	1075
Ghana	8.6	3.0	37	2070,	3	0.8	240	320	100
Greece	8.7	0.9	69	2920,	18	2.0	1000	1040	300
Guatemala	5.0	3.1	—	2000,	9	—	350	140	115
Guinea	4.0	2.2	27	2060,	3	0.7	82	110	280
Haiti	4.3	2.5	45	1930,	5	0.5	90	20	20
Honduras	2.5	3.3	49	2070,	12	1.2	270	120	150
Hong Kong	3.8	2.1	70	2370,	21	3.0	740	1300	1425
Hungary	10.3	0.4	69	3150,	35	3.0	1000	1410	470
India	537.1	2.2	41	1950,	5	2.0	1000	110	8
Indonesia	115.0	2.5	60	1900,	2	1.2	1000	- 17	17
Iran	29.0	3.0	50	2030,	9	1.4	400	245	135
Iraq	9.5	3.3	52	2050,	11	1.7	300	210	170

Table 1. (Continued)

Country	Population (millions)	Approximate population growth (% per year)	Life expectancy at birth (years)	Food (calories per day and % from animal sources)	Education (teachers per 1000 population)	GNP per capita (1970 dollars)	Electric power (kWh per capita per year)	Imports plus exports per capita (1970 dollars)
Ireland	3.0	0.6	71	3460, 41	3.5	1300	2000	900
Israel	2.9	2.7	72	2930, 20	7.3	1800	2350	750
Italy	53.7	0.7	73	3000, 21	7.0	1600	2200	525
Ivory Coast	4.3	2.4	41	2430, 6	0.7	300	130	200
Jamaica	1.8	2.1	64	2280, 17	1.0	600	840	435
Japan	104.7	1.1	72	2450, 14	6.1	1800	3400	365
Jordan	2.3	3.4	52	2400, 7	2.4	270	80	95
Kenya	11.0	3.1	49	2220, 11	0.7	139	45	55
North Korea	13.9	2.8	58	2270, 4	5.1	250	1000	20
South Korea	31.5	2.4	62	2450, 4	2.6	260	325	90
Laos	3.0	2.5	48	2040, 5	0.3	90	5	20
Lebanon	2.7	3.0	—	2400, 13	—	540	430	275
Lestho	1.0	1.8	49	—	0.4	70	8	40
Liberia	1.4	2.2	54	2290, 4	1.0	357	450	315
Libya	2.0	3.7	52	2600, 10	3.4	1900	200	1400
Malagasy Republic (Madagascar)	6.8	2.7	38	2300, 9	1.0	130	30	50
Malawi	4.4	2.5	39	2400, 5	0.2	70	30	30
Malaysia, East	1.6	2.4	—	2430, 10	2.3	300	180	—
Malaysia, West	9.0	2.8	—	2190, 11	2.3	330	370	300
Mali	5.0	2.1	37	2130, 9	0.7	70	8	16
Mauritania	1.2	2.2	41	2360, 8	—	170	60	120
Mexico	48.2	3.2	62	2620, 11	2.7	650	560	80
Mongolia	1.2	2.7	58	2540, 37	—	—	420	150

Morocco	15.0	3.2	51	2130, 9	1.0	200	130	75
Mozambique	8.2	2.1	41	2130, 4	0.3	160	70	60
Nepal	11.4	1.8	41	2030, 6	0.6	85	7	80
Netherlands	12.8	1.2	74	3200, 38	7.8	2200	3100	2000
New Zealand	2.9	1.5	71	3450, 51	—	2190	4700	880
Nicaragua	1.8	3.0	50	2330, 13	1.4	400	310	185
Niger	4.0	2.9	41	2170, 9	0.1	90	10	225
Nigeria	56.0	2.5	37	2300, 5	0.3	99	30	40
Norway	3.9	0.8	73	2960, 33	9.0	2800	15000	6200
Pakistan (1972)	60.	2.4	50	2300, 11	1.4	180	90	30
Panama	1.4	3.0	59	2400, 16	4.0	650	600	310
Paraguay	2.4	3.4	59	2660, 21	3.0	240	90	55
Peru	13.6	3.1	54	2270, 14	3.0	360	440	1000
Philippines	36.7	3.0	41	2020, 13	2.5	259	200	60
Poland	32.6	0.8	70	3140, 30	5.0	880	2000	220
Portugal	9.0	0.7	68	2850, 18	3.5	600	850	265
Puerto Rico	2.7	1.3	72	2530, 33	5.0	2000	3000	3870
Rhodesia, Southern	5.3	3.4	70	2550, 10	0.8	280	1200	130
Romania	20.3	1.1	68	3010, 18	2.5	800	1700	190
Rwanda	3.7	2.9	41	1900, 3	0.2	60	8	15
Saudi Arabia	7.5	2.8	42	2080, 8	0.9	360	100	400
Senegal	3.9	1.7	64	2300, 10	0.5	200	75	90
Sierra Leone	2.5	2.3	41	2160, 4	0.7	170	80	90
Singapore	2.1	2.0	68	2430, 16	4.4	900	1100	2000
Somalia	2.8	2.4	39	1770, 21	—	70	10	27
South Africa	21.4	3.1	49	2800, 20	—	700	2500	285
Spain	34.0	1.1	70	2760, 20	2.9	900	1690	215
Sri Lanka (Ceylon)	12.5	2.3	62	2200, 4	—	170	65	60
Sudan	15.7	2.8	48	2090, 15	0.7	110	25	40
Sweden	8.1	0.7	74	2930, 40	—	4000	7580	1700
Switzerland	6.3	1.2	72	3150, 37	—	3160	5200	1800
Syria	6.3	3.3	53	2450, 11	1.4	260	160	80
Taiwan	14.7	2.3	—	—	4.1	300	770	50

Table 1. (Continued)

Country	Population (millions)	Approximate population growth (% per year)	Life expectancy at birth (years)	Food (calories per day and % from animal sources)	Education (teachers per 1000 population)	GNP per capita (1970 dollars)	Electric power (kWh per capita per year)	Imports plus exports per capita (1970 dollars)
Tanzania	13.3	2.6	41	—	0.2	100	30	40
Thailand	34.4	3.1	—	2210, 8	1.0	190	130	55
Togo	2.0	2.6	35	2210, 3	0.5	140	140	63
Trinidad and Tobago	1.0	1.4	64	2360, 19	1.6	700	1200	1025
Tunisia	5.1	2.6	52	2220, 8	1.7	270	160	100
Turkey	35.6	2.5	54	2760, 10	1.6	350	240	15
Uganda	9.8	2.7	—	2160, 9	0.3	135	80	40
United Kingdom	55.0	0.4	70	3175, 43	7.0	2100	4500	750
United States	204.9	1.0	71	3250, 40	8.5	4700	8000	410
Upper Volta	5.4	2.1	32	2060, 4	—	50	5	12
USSR	242.0	1.1	70	3180, 21	—	1150	3100	101
Uruguay	2.8	1.2	69	2760, 40	4.6	800	800	160
Venezuela	10.3	3.4	64	2400, 18	1200	1000	1200	550
North Vietnam	21.0	2.3	50	2000, 9	—	90	—	—
South Vietnam	18.0	2.1	50	2200, 7	1.0	200	70	—
Yemen (Aden) Peoples Democratic Republic	1.4	2.8	42	2020, 27	0.6	120	30	25
Yemen, Arab Republic of	6.5	2.8	42	1910, 11	—	50	2	50
Yugoslavia	20.3	1.0	67	3150, 19	3.0	500	1200	200
Zaire	21.6	3.9	—	2040, 4	0.4	99	150	60
Zambia	4.2	3.0	—	2250, 6	0.6	386	220	350

* Compiled and computed from several authentic sources. Statistics for many countries are inadequate, and all tabulated values should be considered approximations. Details follow.

Column 1. Many countries did not take a census in 1970. Some approximations are included.
Column 2. Largely 10-year approximations, prior to 1972.
Column 3. In several countries birthrate and mortality statistics are incomplete.
Column 4. Quoted from UN Statistical Yearbook, 1973, Section 162, page 524. Copyright, United Nations. Used with permission.
Column 5. Teachers in upper grades and colleges only; kindergarten and primary omitted. Statistics are not strictly comparable due to variations in school classification, especially as regards vocational and teacher-training schools.

Column 6. Published data on GNP were converted to 1970 constant dollars and averaged per capita by total population in column 1.
Column 7. Totals are gross electricity generated by public utilities. A few totals include output of private plants.
Column 8. Reported total imports plus exports were converted to dollars at published exchange rates, 1970.

Table 2.   U.S. Position in the World Metal Market*

Production, Consumption, and Reserves

Metal	U.S. position, 1960 (% of world) Consumption	Production	Reserves
Aluminum	29	8	1
Cadmium	47	47	12
Chromium	25	2	0
Cobalt	27	5	2
Copper	26	25	15
Gold	6	4	5
Lead	25	11	16
Magnesium	36	39	No limit
Manganese	14	1	0
Mercury	21	14	5
Molybdenum	47	92	61
Nickel	31	4	1
Platinum	70	2	0
Silver	44	14	15
Tin	29	0	0
Titanium	48	34	3
Tungsten	18	10	5
Vanadium	32	72	68
Zinc	28	13	30

* Computed from data of the U.S. Geological Survey.
References:
  Yearbook, American Bureau of Metal Statistics, New York.
  Mineral Economics, National Academy of Sciences, 1969.

Table 3.   World Population and Its Growth, 1900–1985*

UN Data and Projections, World Regions

Table A: Approximate world population (billions)

Year	1900	1910	1920	1930	1940	1950	1955	1960	1965	1970	1975
Population	1.58	1.66	1.81	2.01	2.24	2.49	2.71	2.98	3.30	3.63	4.02

Table B: Population and its growth, world regions

	Population (millions)				Population growth (%)			
	Census		Projected		1960–1970		1975–1985	
Region*	1960	1970	1980	1985	Total, 10 years	Average annual	Total, 10 years	Average annual
World, total	2982	3632	4457	4933	21.8	2.0	22.7	2.1
Developed countries	961	1077	1196	1262	12.1	1.2	11.3	1.1
Developing countries	2020	2555	3260	3672	26.5	2.4	27.1	2.4
Europe (less USSR)	425	462	497	515	8.7	0.8	7.5	0.7
Northern America	199	228	260	281	14.6	1.4	15.6	1.5
USSR	214	243	271	287	13.6	1.3	12.2	1.2
Oceania	16	19	24	27	18.8	1.9	24.3	2.2
Latin America	213	283	377	435	32.9	2.9	33.0	2.9
Africa (less Arab states)	270	344	457	520	27.4	2.5	31.6	2.8
Asia (less Arab states)	1618	2019	2530	2813	24.8	2.2	24.2	2.2
Arab states	92	123	171	202	33.7	2.9	39.3	3.3

* Compiled and computed. Based largely on Unesco Statistical Yearbook 1972, Table 1.2. With permission.

* Regional groups:

Developed countries: all of Europe, USSR, United States, Canada, Japan, Israel, Australia, New Zealand, South Africa.

Developing countries: rest of the world.

Northern America: United States, Canada, Bermuda, Greenland, St. Pierre, Miquelon.

Latin America: rest of America.

Arab states: Algeria, Egypt, Libya, Morocco, Sudan, Tunisia, Bahrain, Iraq, Jordan, Kuwait, Lebanon, Qatar, Syria, Saudi Arabia, Peoples Democratic Republic of Yemen, Yemen.

Table 4. U.S. Population and Its Growth, 1800–2000*

Total Growth and Average Yearly Growth

		Growth in previous period				Growth in previous period	
Year	Total population (millions)	Total (%)	Average annual (%)	Year	Total population (millions)	Total (%)	Average annual (%)
1800	5.3	—	—	1965	194.3	8	1.5
1820	9.6	81	3.0	1970	204.9	5	1.0
1840	17.1	78	2.8	1971	207.0	1.0	1.0
1860	31.5	84	3.1	1972	208.8	0.9	0.9
1880	50.2	60	2.4	1973	210.4	0.8	0.8
1900	76.1	52	2.2	1974	211.9	0.7	0.7
1910	92.4	21	2.0	1975	213.4	0.7	0.7
1920	106.5	15	1.4	1980 min.	221.9	8.3	0.8
1930	123.1	16	1.5	1980 max.	231.0	12.7	1.1
1940	132.6	8	0.7	1990 min.	239.1	7.8	0.7
1950	152.3	15	1.4	1990 max.	266.2	15.2	1.4
1955	165.9	9	1.7	2000 min.	250.7	4.9	0.5
1960	180.7	9	1.7	2000 max.	300.4	12.8	1.1

* Condensed and computed from the statistics of the Bureau of the Census (July 1 data).

For population projections for 1975–2000 four estimates were actually made, based on four family sizes (average number of births to completion of child bearing per 1000 women was the criterion), "moving gradually toward" 1800, 2100, 2500, 2800. In the above table the lowest and the highest are shown, that is, an eventual average of 1.8 children per female in the population to 2.8 children per female in the population. A net immigration of 400,000 was included.

The projections are periodically revised, and in 1975 the preferred range of eventual estimated births per female was changed from 1.8 to 2.8 to 1.7 to 2.7. This would change the maximum estimated population in the year 2000 from 300.4 million to 287.0 million (last line). Long-range fertility estimates cannot be made with accuracy.

References:

Statistical Abstract of the United States, 1974 and 1975, Department of Commerce, Bureau of the Census.

Table 5. U.S. Economy by Decades, 1900–1973*

Personal Income, Energy, Cars, Education

Totals:

Year	Population (millions)	GNP (dollars, billions)	Electric power (kWh, billions)†	Motor vehicles (millions)	College enrollments (millions)
1973	210.4	1272	1718	125.2	8.3
1970	204.9	977	1640	109.0	7.5
1960	180.7	504	842	73.80	3.6
1950	152.3	285	389	49.16	2.7
1940	132.6	100	180	32.45	1.5
1930	123.1	91	115	26.75	1.1
1920	106.5	89	57	9.24	0.6
1910	92.4	32	20‡	0.47	0.4
1900	76.1	18	5‡	0.01	0.25

Per capita:

Year	Personal income per capita before taxes — Dollars reported	Personal income per capita before taxes — In 1970 dollars	Electric power, utilities (kWh per capita per year)	Motor vehicles per 100 population	College students per 1000 population
1973	4920	3985	8165	59.5	39.4
1970	3945	3945	8005	53.2	36.6
1960	2220	2575	4660	40.8	20.0
1950	1500	2025	2554	32.3	17.7
1940	594	1622	1357	24.5	11.3
1930	625	1563	934†	21.7	9.0
1920	689	965	535†	8.8	5.6
1910	300‡	(930)	151†	0.5	4.3
1900	200‡	(776)	38†	0.01	3.3

* Sources of data: Bureau of the Census, Department of Commerce, FPC, Federal Highway Administration, Office of Education. (See also Table 28.)
† It should be noted that in early years most factories and large buildings generated their own power.
‡ Approximately.

Table 6.   U.S. Crops of Woods and Fibers, 1950–1973*

Land Use, Production, and Export

	1950	1960	1965	1970	1973
Wood					
Total timber production (cubic feet, billions)	10.8	10.2	11.5	11.7	12.3
Lumber production, softwoods					
(board feet, billions)	30.6	26.7	29.3	27.5	31.6
Lumber consumption, softwoods					
(board feet, billions)	34.3	29.0	32.7	31.9	38.5
Lumber production, hardwoods					
(board feet, billions)	7.4	6.3	7.5	7.1	7.0
Lumber consumption, hardwoods					
(board feet, billions)	7.4	6.3	8.1	6.5	7.5
Timber used, production (cubic feet, billions)					
Saw logs, lumber	5.9	5.1	5.7	5.4	5.7
Veneer logs, plywood	0.4	0.7	1.0	1.1	1.3
Pulpwood	1.5	2.6	3.1	3.8	3.7
Fuel wood	2.3	1.3	0.9	0.5	0.5
Paper and paperboard					
Consumption (tons, millions)	29.0	39.2	49.1	58.0	67.2
Wastepaper salvage (%)	27.	23.	21.	20.	20.
Cotton					
Acreage harvested (millions)	—	15.3	13.6	11.1	12.0
Production (bales, millions)	—	14.	15.	10.	13.
Exports (% of production)	—	49.	27.	30.	38.
Fibers, wool, yarn: Exports less imports					
(dollars, millions)					
Fibers, cotton, wool, hair, yarn					
(fabrics not included)	—	1500	−32	485	1840

* Compiled from data published by the Department of Agriculture.

Total forests, 755 million acres; commercial timberland, 500 million acres, owned and managed as follows: federal, 22%; state and municipal, 6%; private, 72%.

Table 7. Food Supplies, Affluence versus Poverty, 1970*

Upper and Lower 20% of the Countries of the World

Country	Population (millions)	Food supply Calories per day per capita	% Animal origin
Ten affluent countries			
East Germany	17.3	3040†	37
Australia	12.6	3160†	45
United Kingdom	55.0	3175	43
USSR	242.8	3180†	21
West Germany	59.6	3180	42
Netherlands	13.0	3200	38
Canada	21.4	3200†	46
Belgium	9.7	3230	39
France	50.8	3270	41
United States	204.9	3300	40
Total population	687.1		
Ten poorer countries			
Bolivia	4.9	1760†	14
Somalia	2.8	1770†	21
El Salvador	3.5	1850	10
Algeria	14.0	1890†	6
Rwanda	3.6	1900	3
Indonesia	121.2	1900	2
Ethiopia	25.0	1980	9
India	537.1	1990	5
Philippines	38.5	2040	13
Ghana	9.0	2070†	3
Total population	759.6		

* Compiled from UN sources, see Table 1.
† Data earlier than 1969–1970.

Table 8. U.S. Food Production, Consumption, and Trade, 1950–1973*

Acreage, Employment, and Market Value in 1970 dollars.
Field, Animal, and Fish Products

	1950	1955 (1954)	1960	1965	1970	1973
Total land area (acres, billions)	2.27	2.27	2.27	2.27	2.26	—
Cropland (%)	18.0	—	17.3	17.1	17.0	—
Pasture, grazing (%)	45.0	—	41.5	40.6	39.3	—
Employment, farm only (millions)	9.93	8.38	7.06	5.61	4.52	4.34
Food manufacture (millions)	1.79	—	1.79	1.76	1.78	1.74
Food stores, retail (millions)	—	(1.44)	1.36	1.47	1.73	1.89
Average farm size (acres)	213	258	297	340	373	383
Agricultural exports (dollars, billions)‡	4.8†	6.2†	5.7†	7.10	7.36	14.32
Production, nine principal food crops (dollars, billions)	—	14.7†	13.97	16.19	16.51	31.78
Corn, oats, sorghums, hay (dollars, billions)	—	7.6	8.76	10.16	9.96	16.98
Wheat (dollars, billions)	2.3	2.2	2.74	2.02	1.80	5.27
Soybeans (dollars, billions)	0.7	0.9	1.38	2.45	3.22	7.17
Rice (dollars, billions)	0.23	0.25	0.25	0.43	0.43	1.06
Potatoes, all (dollars, billions)	0.43	0.48	0.60	0.83	0.72	0.97
Sugar beets (dollars, billions)	0.13	0.15	0.22	0.29	0.39	0.34
Animal products, total (dollars, billions)	19.4	17.4	18.6	20.3	23.5	32.2
Beef (dollars, billions)	6.95	5.66	7.01	7.75	11.40	16.14
Hogs and sheep (dollars, billions)	4.85	4.10	3.64	4.28	5.14	6.41
Poultry and eggs (dollars, billions)	3.47	2.88	2.73	2.70	2.83	3.26
Milk (dollars, billions)	4.09	4.75	5.17	5.54	4.16	6.35
Farmer's share of consumer's dollar (%)	47	41	39	40	39	46
Fish, domestic production (tons, millions)	—	2.41	2.47	2.39	2.46	2.37
Fish, imports (tons, millions)	—	1.16	1.64	2.88	3.28	2.76
Fish, for human food (%)	—	55	52	49	54	69

* Complied and computed from U.S. government reports.
  Since 1970 the yearly agricultural exports have been much larger than imports.
† Approximate only.
‡ Exports of all agricultural products. Includes Puerto Rico. Forest products excluded. For food exports see Table 21.

Table 9.   Youthful World Populations, 1950–1970*

Populations Less Than 25 Years of Age (millions)

World and subdivisions†	Year	Population 0 to 25 years of age	% of total population
United States only, data	1950	63.3	41.6
from Bureau of	1960	80.7	44.6
the Census	1970	94.4	46.0
World total, UN data	1960	1637	54.9
	1965	1805	54.8
	1970	2010	55.3
Developed countries	1960	423	44.0
	1970	470	43.7
Developing countries	1960	1213	60.0
	1970	1540	60.3
Europe	1960	172	40.5
	1970	187	40.6
Northern America	1960	89	45.0
	1970	107	47.0
USSR	1960	101	46.9
	1970	109	45.0
Oceania	1960	7.5	47.9
	1970	9.7	49.9
Africa (less Arab states)	1960	127	62.2
	1970	169	65.4
Asia (less Arab states)	1960	953	58.9
	1970	1186	58.7
Arab states	1960	58	62.7
	1970	78	63.7
Latin America	1960	129	60.5
	1970	173	61.2

* Compiled and computed from UNESCO Statistical Yearbook 1972, Table 1.2, p. 27, with permission.

† For countries included in each region, see Table 3.

Table 10. Some Annual Growth Rates in the United States, 1950–1973*

Population, GNP, Personal Income, Energy

	Compound yearly growth (%)			
	1950–1960	1960–1965	1965–1970	1970–1973
Total population	1.7	1.4	1.1	0.9
GNP (constant dollars)†	3.3	4.8	3.2	5.0
Personal income per capita (constant dollars)†	1.8	3.4	2.8	4.6
Total energy consumption (Btu)‡	2.7	3.7	4.6	4.0
Petroleum consumption (Btu)	3.9	3.1	6.4	5.5
Natural gas consumption (Btu)	7.5	5.0	6.2	2.3
Coal consumption (Btu)	−2.1	3.4	0.6	0.7
Electrical consumption (kWh)§	9.3	6.9	7.9	7.0
GNP per capita†	1.5	3.2	2.1	3.9
Total energy per capita	1.0	2.2	3.5	2.9
Petroleum per capita	2.1	2.1	5.3	4.4
Natural gas per capita	5.7	3.5	5.1	1.1
Electric power per capita	7.5	5.3	6.7	5.8

* Compiled and computed from federal publications: DOI, FPC, Library of Congress, Department of Commerce, BEA. Growth values are based on listed years only.
† Corrections for inflation as recommended by the BEA.
‡ Energy data based on Tables 12 and 17.
§ Electrical utilities' sales per the FPC.

Table 11. Distribution of U.S. Fuel Use, 1965–1974*

Percentages Supplied to the Consuming Sectors

	1965	1970	1972	1973	1974†
Gross energy input, all purposes (Btu × 10⁻¹⁵)	53.4	67.1	72.0	74 7	73.1
Industrial (%)‡	32.3	30.2	28.5	28.7	29.0
Transportation (including military) (%)	23.8	24.5	25.1	25.3	25.0
Residential and commercial (%)	22.2	20.8	20.3	19.1	18.9
Electrical utility generation (%)	20.7	24.2	25.8	26.5	26.8
Coal (%)	54.5	46.2	44.1	45.7	44.6
Oil (%)	6.1	11.9	15.6	16.9	16.1
Gas (%)	21.0	24.3	21.5	18.3	17.2
Hydroelectric (%)	18.4	16.2	15.6	14.6	16.1
Nuclear (%)	—	1.4	3.1	4.5	6.0

* Data from the FPC and Bureau of Mines.

    Energy inputs to consuming sectors are *fuel* use only and do not include electric energy purchased from utilities.

† Preliminary.

‡ Percentages of gross energy input.

Table 12. U.S. Energy Consumption and Imports, 1950–1972*

Total and per Capita Use and Distribution

	1950	1960	1965	1970	1972
Total energy input, all sources (Btu $\times$ 10^{-15})†	34.1	44.8	53.4	67.1	72.0
Source distribution (%)‡					
Coal, all grades	38.0	22.8	22.6	18.9	17.5
Petroleum and imports	37.3	41.7	40.0	42.5	43.8
Natural gas and liquids	20.5	31.8	33.5	34.2	33.7
Hydroelectric (utility and other)	4.2	3.7	3.9	3.8	4.0
Nuclear power	—	—	—	0.6	1.0
Net imports (% of consumption of each fuel)					
Petroleum	9.6	19.0	23.0	23.2	37.7
Natural gas	—	1.2	2.8	3.5	4.2
Consumption per capita					
Coal (Btu $\times$ 10^{-6})	85	58	64	62	60
Petroleum (Btu $\times$ 10^{-6})	84	103	111	145	158
Natural gas (Btu $\times$ 10^{-6})	41	71	83	108	111

* Complied and computed from data published by the Library of Congress, *Energy Facts*; Bureau of Mines, *Minerals Yearbook* and *Mineral Industry Survey*. Gross energy consumption, all purposes.

† These energy inputs include nonfuel uses and *all losses,* that is, total input as delivered from mine, well, or port.

‡ Percentage of total energy input supplied by each fuel; approximate only.

Table 13. World Fuel Reserves, 1972*

Various Guesses on Recoverable Fuel Underground

Table A: Weighted Estimates of reserves

Country or region	Natural gas	Petroleum	Coal, all grades
United States, total (commercial units)†	270	42	500
United States (Btu $\times$ 10^{-15})	280	235	12,500
United States (kWh $\times$ 10^{-12})	81.5	69	3,660
World (commercial units)†	1,800	640	2,550
World (kWh $\times$ 10^{-12})	543	1,050	187,500
	Approximate percentage shares of world total		
United States, total	15.0	6.6	19.6
Total, western hemisphere	22.2	12.5	21.6
Total, eastern hemisphere	77.8	87.5	78.4
Western Europe	10	2	8
Eastern Europe and USSR	33	14	31
Middle East	25	56	0
Other Africa	3	10	0
Other Asia	7	5	39

Table B: Consumption and life of fuels, United States and world, 1972— Static life of proved reserves at present rates of use

	Natural gas (cubic feet $\times$ 10^{12})	Petroleum (barrels $\times$ 10^9)	Coal (tons $\times$ 10^9)
U.S. consumption (Table 12)	27.7	4.0	0.672
U.S. reserves	270	42	500
Apparent static life, United States, (years)	10	11	750
World consumption	42	18	3.3
World reserves	1800	640	2550
Apparent static life, world, (years)	43	36	775

* This table was compiled and computed from available basic sources including publications of the federal government, trade associations, UN, OECD, and so on. In cases of disagreement, weighted approximations were made by the author.

In the final analysis, all published data on fuel reserves are estimates, and there are wide disagreements among the authorities. In most fields the total underground deposits are two or three times the proved reserves, but as techniques are improved a higher fraction of the total can be extracted. Higher fuel prices will also make profitable a higher percentage extraction and more new discoveries. Hence the static life is only a very rough indicator, and one that changes every year.

† Commercial units: gas, cubic feet $\times$ 10^{12}; oil, barrels $\times$ 10^9; coal, tons $\times$ 10^9.

Table 14. Electrical Utility Power, 1950–1974*

Capacities and Outputs from U.S. Fuel, Hydroelectric, and Nuclear Plants

	Units	1950	1960	1970	1972	1974†
Electricity generated, utility power for public use, industrial plants not included						
All utilities, total for year	kWh × 10⁻¹²	0.33	0.75	1.53	1.75	1.87
From fuel	%	70.8	80.6	82.4	81.3	77.9
From hydroelectric, utility only	%	29.2	19.3	16.2	15.6	16.1
From nuclear reactors	%	0	0.1	1.4	3.1	6.0
Existing capacity, utility only						
All utilities, total	kW × 10⁻⁶	69	168	341	400	47.6
Fuel stations	%	73.9	80.8	81.7	82.0	80.2
Hydroelectric stations	%	26.1	19.0	16.1	14.3	13.4
Nuclear stations	%	0	0.2	2.2	3.7	6.4
Nonfuel capacity						
Developed hydroelectric (utility and other)	kW × 10⁻⁶	18.7	33.2	52.0	53.8	—
New England and Atlantic states	%	31.	21.	21.	21.	—
Central states	%	25.	22.	21.	20.	—
Mountain and Pacific states	%	44.	57.	58.	59.	—
Undeveloped hydroelectric	kW × 10⁻⁶	88.	114.	128.	126.	—
New England and Atlantic	%	20.	16.	14.	13.	—
Central states	%	19.	16.	10.	10.	—
Mountain and Pacific	%	61.	68.	76.	77.	—
Nuclear, in operation, 1974	kW × 10⁻⁶	—	—	—	—	29.
Nuclear, under construction, 1974	kW × 10⁻⁶	—	—	—	—	58.†
Nuclear, on order, 1974	kW × 10⁻⁶	—	—	—	—	116.‡

* Compiled from data published by the FPC, AEC, and Bureau of Reclamation.
† 1974 data not finalized.
‡ Some postponements or cancelations, 1974–1975.

Table 15.   Solar Heat Available in U.S. Cities*†

Maximum Heat Quantities per Square Foot

	Seasonal averages			
Typical zones, north latitude 32° and 40°, normal irradiation	Dec. to Feb.	Mar. to May	June to Aug.	Sept. to Nov.
32° Zone‡				
Outdoors, average (Btu per day)	2855	3395	3275	2935
Outdoors, 24-hour average (Btu per hour)	119	142	136	122
Within horizontal collector (Btu per day)	1090	1995	2140	1350
Within horizontal collector (Btu per hour)	45	83	89	56
40° Zone§				
Outdoors, average (Btu per day)	2540	3350	3320	2715
Outdoors, 24-hour average (Btu per hour)	106	140	138	113
Within horizontal collector (Btu per day)	785	1875	2095	1085
Within horizontal collector (Btu per hour)	33	78	87	45

* Compiled and computed.

† Among the U.S. cities that average 110 or more days per year of clear sunshine are the following: Norfolk, Va.; Memphis, Tenn.; New Orleans, La.; Jackson, Miss.; Kansas City, Mo.; Dallas, Texas; Oklahoma City, Okla.; San Antonio, Texas; Dodge City, Kans.; Denver, Colo.; Albuquerque, N. Mex.; Phoenix, Ariz.; San Diego, Calif.; Fresno, Calif.; San Francisco, Calif.

Among the U.S. cities that average 75 or *fewer* days of clear weather per year are the following: Burlington, Vt.; Albany, N.Y.; Miami, Fla.; Rochester, N.Y.; Buffalo, N.Y.; Pittsburgh, Pa.; Charleston, W. Va.; Asheville, N.C.; Cleveland, Ohio; Columbus, Ohio; Indianapolis, Ind.; Duluth, Minn.; Seattle, Wash.; Portland, Ore.; Honolulu, Hawaii.

‡ Typical cities, 31° to 33°N latitude: Charleston, S.C.; Savannah, Ga.; Montgomery, Ala.; Jackson, Miss.; Shreveport, La.; Dallas, Texas; Carlsbad, N. Mex.; Tucson, Ariz.; San Diego, Calif.

§ Typical cities, 39° to 41°N latitude: Trenton, N.J.; Harrisburg, Pa.; Columbus, Ohio; Indianapolis, Ind.; Springfield, Ill.; Lincoln, Neb.; Denver, Colo.; Provo, Utah; Redding, Calif.

Heat quantities are approximate direct solar heat in an average location on a typical cloudless day. Diffuse sky radiation and reflected radiation are not included.

Table 16.   Energy Projections for 1980*

Energy Self-Sufficiency—Prospects and Results

Source of energy	Equivalent millions of barrels per day, at three prices for oil		
	$7 per barrel	$9 per barrel	$11 per barrel
Estimates of total demand, 1980			
All sources	44.2–45.6	42.4–45.6	40.6–45.6
Estimates of supply, 1980			
Oil and gas liquids	10.5–13.6	12.6–13.6	13.6–14.8
Natural gas	11.5–14.7	11.5–15.8	11.5–16.9
Coal	7.1	8.0	8.0
Nuclear and hydro	6.2	6.2	6.2
Total energy supply	38.4–38.5	39.3–42.6	39.4–46.0
Total deficiency	5.8–7.2	−0.4–6.3	−5.4–6.2

* A study by the MIT Energy Laboratory, 1973–1974. Data from *Technology Review,* **76**(6) 23–58 (May 1974), with permission.

All energy quantities converted to equivalent millions of barrels per day of crude oil. A range of estimated results is given if more than one estimate was reported. Prices are in 1973 constant dollars.

Analyses were based on econometric models and on judgmental forecasts, with differing results in the ranges indicated.

Contributions from nuclear and hydroelectric power were assumed to be unaffected by fuel prices, and contributions from new energy sources were assumed negligible.

In the judgmental forecasts of demand, electrical utilities were expected to use one-third of the entire energy supply in 1980.

A short-term program to achieve energy independence was recognized as very costly and calling for new government energy policies in the fields of: (1) federal regulation, (2) actions to let the market work effectively, (3) actions to redress adverse effects on income distribution, (4) actions to provide security against oil import disruption.

For comparisons see Tables 11, 12, 18, 19.

Table 17. The U.S. Energy-Supply Industries, 1950–1973*

Output, Employment, and Finances

	1950	1960	1965	1970	1972	1973
Petroleum production (barrels, millions)	1,974	2,575	2,849	3,517	3,455	3,361
Petroleum net imports (barrels, millions)	199	591	833	1,153	1,654	2,200
Gas production (cubic feet, billions)	6,282	12,771	16,040	21,921	22,532	22,648
Gas, net imports (cubic feet, billions)	26	144	430	751	941	956
Coal production (tons, millions)	560	434	527	613	602	599
Coal net exports (tons, millions)	29	38	52	72	57	53
Employment, oil and gas, extraction (thousands)	518	511	437	462	—	—
Employment, gas utilities (thousands)	—	155	154	161	163‡	161
Employment, coal mining (thousands)	483	190	149	144	130	134
Employment, electric utilities (thousands)	—	253	253	289	300‡	311
Construction, oil processing (dollars)†	1.63	2.89	4.03	5.62	5.25	—
Construction, oil pipelines (dollars)	—	0.13	0.12	0.29	0.28	0.40
Construction, gas utilities (dollars)	1.2	1.85	1.94	2.51	2.82	2.97
Construction, electrical utilities (dollars)	1.57	2.03	2.59	5.81	7.58	8.33
Plant investment, gas utilities (dollars)	7.6	20.8	28.2	38.5	43.2	45.6
Plant investment, electrical utilities (dollars)	19.0	45.5	59.7	93.3	116.6	130.8
Operating revenue, gas utilities (dollars)	2.55	8.70	11.52	16.38	19.75	20.59
Operating revenue, electrical utilies (dollars)	5.1	11.5	15.2	22.1	27.9	31.7
Operating net income, gas utilities (dollars)	0.32	0.83	1.11	1.43	1.79	1.98
Operating net income, electrical utilities (dollars)	0.82	1.78	2.58	3.41	4.42	4.99

* Sources: BEA, Bureau of Mines, Bureau of the Census, *Construction Reports,* Bureau of Labor Statistics, Library of Congress, *Energy Facts,* FPC, American Gas Association, Edison Electric Institute.
† Dollar values are billions of current dollars.
‡ Approximate only.

Table 18.   One Forecast of U.S. Energy Consumption to the Year 2000*

Energy Sources and Uses (Btu × 10^{-15})

	1970	1975	1980	1985	2000
Total gross energy input	67.1	80.3	96.0	116.6	191.9
Conversion losses (%)	11.	15.	20.	27.	52.
Net energy consumption, at point of use†	56.2	65.1	76.1	89.7	140.1
Assumed yearly growth, previous 5 years					
(%)	—	3.0	3.1	3.3	3.0
Yearly growth per capita (%)‡	—	2.0	2.1	2.3	2.0
Energy consumption sectors, at point of use†					
Residential and commercial	17.0	20.2	23.9	27.7	39.6
Fraction of total (%)	25.	25.	25.	24.	21.
Assumed yearly growth (%)	—	3.5	3.5	3.0	2.4
Industrial	22.5	25.9	29.4	34.9	57.8
Fraction of total (%)	34.	32.	31.	30.	30.
Assumed yearly growth (%)	—	2.9	2.6	3.5	3.5
Transportation	16.5	19.1	22.9	27.1	42.7
Fraction of total (%)	25.	24.	24.	23.	22.
Assumed yearly growth (%)	—	3.0	3.8	3.4	3.1
Energy supply sectors (% of gross energy input)					
Coal	12.7	13.8	16.1	21.5	31.4
Petroleum	29.5	35.1	42.2	50.7	71.4
Supplemental, imports and synthetics (%)	7.4	13.0	18.4	27.1	50.2
Natural gas	22.0	25.2	27.0	28.4	34.0
Supplemental, imports (%)	0.8	2.6	4.0	5.9	11.1
Nuclear power	0.2	2.6	6.7	11.8	49.2
Hydroelectric power	2.6	3.6	4.0	4.3	6.0

* From *Statistical Abstract of the United States* Table 881, 1975. Original sources: *Minerals Yearbook, U.S. Energy through the Year 2000, Mineral Industry Survey.*
† Net energy consumption, including nonfuel uses, equals gross energy input less conversion losses. See Table 11 for percentages of gross energy input.
‡ Average population growth assumed 1% per year.

Table 19. Growth to the Year 2000—Preferred Assumptions*

Historical Growth and Realistic Future Slow Growth

	Average yearly growth rates		Projection estimates at realistic future slow growth rates†				
	Boom times (%)	Future slow growth (%)	1970	1980	1985	1990	2000
World							
Population (billions)[a]	2.0	1.75	3.63	4.32	4.71	5.13	6.11
Food supply (%)	—	2.5	100	128	145	164	210
United States:							
Population (millions)[b]	1.2	0.8	204.9	221.9	231.0	240.3	260.2
Farm output, foods (%)[c]	2.0	3.0	100	134	156	181	243
GNP (1970 dollars, billions)[d]	4.5	2.5	977	1251	1415	1600	2050
Total energy input (Btu × 10^{-15})[e]	4.0	2.5	67.1	85.9	97.2	100	141
Petroleum (Btu × 10^{-15})[d]	4.5	2.0	29.5	36.0	39.7	43.8	53.4
Gas (Btu × 10^{-15})[d]	6.0	2.0	22.0	26.8	29.6	32.7	39.8
Coal (Btu × 10^{-15})[d]	—	4.5	12.7	19.7	24.6	30.6	47.6
Electric power generated (kWh × 10^{-12})[f]	7.5	4.5	1.53	2.38	2.96	3.69	4.96
Motor fuels (gallons, billions)[g]	5.5	2.5	92.3	118	134	151	194
Airplane fuel (gallons, billions)[h]	—	2.0	7.80	9.51	10.5	11.6	14.1

* Compiled. See footnotes for sources.

† Future values of each item at the realistic slow growth rates selected by the author and recommended as average growth rates 1970–2000, for all projection estimating.

[a] See Table 3.

[b] See Table 4.

[c] See Table 8.

[d] See Table 10.

[e] See Table 12 (losses included).

[f] See Table 14.

[g] Cars, buses, trucks, and combinations; from Federal Highway Administration.

[h] Scheduled air carriers only; from Federal Aviation Administration.

Table 20. World Food Production per Capita, 1961–1972*

Index Numbers Based on 1963 = 100

Year	World population	Food production per capita			
		World	Latin America	Africa	Far East Asia†
1961	96	99	101	95	100
1962	98	100	99	98	100
1963	100	100	100	100	100
1964	102	102	100	98	101
1965	104	101	102	97	96
1966	107	103	100	95	95
1967	109	105	103	99	98
1968	111	107	101	99	101
1969	114	106	102	100	102
1970	116	106	103	100	104
1971	119	108	102	102	102
1972	122	106	100	101	97

* As reported by the FAO. From UN Statistical Yearbook 1973, Sec. 7, with permission.

Conclusion: Even in the face of a rapid increase in world population, world food production per capita has increased very slightly. There was almost *no* increase of the low nutrition level in the underdeveloped countries from 1963 to 1972 (see Table 7 for nutrition levels).

† China excluded.

Table 21.   Exports and Imports of Food, 1955–1974*

Changes in Net Foreign Trade of the United States

	1955	1960	1965	1970	1971	1972	1973	1974
Wheat, exports	483	1,029	1,184	1,112	1,090	1,452	4,151	4,589
Rice, exports		151	244	314	257	389	541	852
Other grains, exports		518	1,135	1,072	980	1,532	3,558	4,673
Animal feed and miscellaneous, exports		87	278	497	551	596	1,266	1,237
Soybeans		336	650	1,216	1,327	1,508	2,757	3,537
Meat and preparations, exports	70	115	162	175	192	252	444	381
Meat and preparations, imports		314	426	1,014	1,050	1,223	1,668	1,344
Meat and preparations, net exports		−199	−264	−839	−858	−971	−1,224	−963
Fruits and nuts, exports	161	265	339	406	430	526	662	758
Fruits and nuts, imports	201	218	339	447	460	496	576	628
Fruits and nuts, net exports	−40	47	0	−41	−30	30	86	129
Vegetables, exports	104	133	148	178	182	209	307	391
Vegetables, imports	45	96	139	289	287	350	409	387
Vegetables, net exports	59	37	9	−111	−105	−141	−102	4
Cattle, imports	−14	62	98	111	107	152	192	107
Fish, imports	214	308	469	794	879	1,205	1,387	1,499
Sugar, imports	414	507	443	725	764	832	918	2,256
Coffee, imports	1,357	1,003	1,058	1,160	1,167	1,182	1,566	1,504
Tea, imports	65	56	57	53	71	63	69	79
Cocoa or beans, imports	—	143	120	201	181	151	212	316
Total food, all exports†	—	2,684	4,002	4,356	4,367	5,660	11,931	13,983
Total food, all imports	—	2,996	3,460	5,375	5,529	6,370	7,986	9,379
Total food, net exports†	—	−312	542	−1,019	−1162	−710	3,945	4,604
Total food (% of total exports)†	—	13.2	14.7	10.2	10.0	11.6	17.0	14.4
Total food (% of total imports)	—	19.9	16.1	13.5	12.1	11.5	11.6	9.3

* Bureau of the Census, and Bureau of International Policy and Research.

    In millions of current dollars. Some quantities are only approximate because of changes in classifications. Values are at point of sale and do not include freight and insurance. Totals include dairy foods and miscellaneous foods, but not alcoholic beverages.

† Soybeans not included in food export totals.

Table 22. Countries with High Population Growth*

Rates of Annual Growth, 10-Year Average
Data for sixty countries, total 1,406 million.

Country	Location	Yearly growth, (%)	Population, 1970 (millions)
1. Annual growth 3.0 to 3.9%, 322 million			
Zaire	Africa	3.9	21.6
Costa Rica	N. Am.	3.7	1.7
Libya	Africa	3.7	2.0
Algeria	Africa	3.5	14.0
El Salvador	N. Am.	3.4	3.4
Jordan	Asia	3.4	2.3
Ecuador	S. Am.	3.4	6.0
Paraguay	S. Am.	3.4	2.4
Venezuela	S. Am.	3.4	10.3
S. Rhodesia	Africa	3.4	5.3
Syria	Asia	3.3	6.3
Iraq	Asia	3.3	9.5
Mexico	N. Am.	3.2	48.2
Colombia	S. Am.	3.2	21.0
S. Africa	Africa	3.1	21.4
Kenya	Africa	3.1	11.0
Thailand	Asia	3.1	34.4
Guatemala	N. Am.	3.1	5.0
Peru	S. Am	3.1	13.6
Iran	Asia	3.0	29.0
Dominican Rep.	N. Am.	3.0	4.3
Philippines	N. Am.	3.0	36.7
Zambia	Africa	3.0	4.2
Yemen	Africa	3.0	7.0
Panama	N. Am.	3.0	1.4
2. Annual growth 2.5 to 2.9%, 435 million			
Ghana	Africa	2.9	8.6
Brazil	S. Am.	2.9	93.2
Albania	Europe	2.9	2.0
Niger	Africa	2.9	4.0
N. Korea	Asia	2.8	13.9
Saudi Arabia	Asia	2.8	7.5
W. Malaysia	Asia	2.8	9.0
Sudan	Asia	2.8	15.7
Uganda	Africa	2.7	9.8
Mongolia	Asia	2.7	1.2
Israel	Asia	2.7	2.9
Tanzania	Asia	2.6	13.3
Togo	Asia	2.6	2.0
Bolivia	S. Am.	2.6	4.9

Table 22. (Continued)

Country	Location	Yearly growth, (%)	Population, 1970 (millions)
Laos	Asia	2.5	3.0
Egypt	Africa	2.5	33.0
Indonesia	Asia	2.5	115.0
Nigeria	Asia	2.5	56.0
Haiti	N. Am.	2.5	4.3
Turkey	Asia	2.5	35.6
3. Annual Growth 2–2.4%, 650 Million			
Ivory Coast	Africa	2.4	4.3
S. Korea	Asia	2.4	31.5
Somalia	Africa	2.4	2.8
E. Malaysia	Asia	2.4	1.6
Afghanistan	Asia	2.3	17.1
Sri Lanka	Asia	2.3	12.5
N. Vietnam	Asia	2.3	21.0
Cuba	N. Am.	2.2	8.4
Guinea	Africa	2.2	4.0
Cent. Afr. Rep.	Africa	2.2	1.6
Mauritania	Africa	2.2	1.2
India	Asia	2.2	536.0
Mali	Africa	2.1	5.0
Singapore	Asia	2.0	2.1

* Source: Statistics per Table 1.

Table 23.   Average National Income per Capita, 1970*

Samples from Each End of the Spectrum

A. High yearly income: Twelve countries, representing about 15% of the
world's population. Average national income per capita, at market prices
(1970 dollars)

Country	Population, 1970 (millions)	Annual income per capita (dollars)
Australia	13.	2629
Belgium	9.7	2413
Canada	21.4	3246
Denmark	4.9	2902
France	50.8	2606
West Germany	60.	2711
Japan	104.6	1658
Netherlands	13.0	2211
Norway	3.9	2538
Sweden	8.0	3730
United Kingdom	55.	1980
United States	205.	4294

For these 549 million people, the population-weighted averages for some other measures
of living standards (1970) were about as follows:

Food (calories per day)	2922.
Education (teachers per 10,000 population)†	74.
Electric power (kWh per capita per year)	6210.
Life expectancy at birth (years)	71.6

Table 23. (Continued)

B. Low yearly income: Thirty-four countries, representing about 15% of world population, and India, representing another 15% of the world's population

Country	Popula-tion, 1970 (millions)	Annual income per capita (dollars)	Country	Popula-tion, 1970 (millions)	Annual income per capita (dollars)
Afghanistan	17.1	57	Mauritania	1.2	160
Bolivia	4.9	190	Mozambique	8.2	145
Burma	27.6	75	Nepal	11.0	85
Burundi	3.6	44	Niger	4.1	92
Central African Republic	1.5	98	Nigeria	55.1	83
Chad	3.7	58	Pakistan	64.	132
Congo	1.2	180	Rwanda	3.6	56
Egypt	33.3	197	Senegal	3.9	190
Ethiopia	25.0	62	Somalia	2.8	62
Guinea	3.9	77	Sudan	15.7	103
Haiti	4.9	87	Thailand	35.8	176
Indonesia	121.2	89	Uganda	9.8	105
Kenya	10.9	131	Upper Volta	5.4	58
Lestho	1.0	95	South Vietnam	17.9	163
Malagasy Republic	6.8	125	Yemen (Arab)	7.0	53
Malawi	4.5	70	Yemen (Aden)	1.4	153
Mali	5.0	71	Zaire	21.6	87
			India	540.	86

For these 544 million people, *plus* the 540 million in India, the population-weighted averages for some other measures of living standards (1970) were about as follows:

Food (calories per day)	2065.
Education (teachers per 10,000 population)‡	8.
Electric power (kWh per capita per year)	88.
Life expectancy at birth (years)	42.§

* Compiled and computed; see Table 1.
† Primary excluded. Data only 90% complete.
‡ Primary excluded. Data only 70% complete.
§ Rough average; data incomplete.

Table 24. Health Services—Affluence versus Poverty*

Doctors, Nurses, and Hospital Beds per 10,000 Population

	Doctors plus dentists	Nurses	Hospital beds
Ten affluent countries total population 687 million			
East Germany	21	†	109
Australia	16	71	123
United Kingdom	18	40	107
USSR	28	43	110
West Germany	23	26	116
Netherlands	16	†	120
Canada	19	66	99
Belgium	18	†	83
France	14†	†	106
United States	21	53	76
Ten poorer countries total population 760 million			
Bolivia	6	3.3	19
Somalia	†	3.2	17
El Salvador	4	10	18
Algeria	1.4	3.3	28
Indonesia	0.4	1.2	7
Rwanda	0.2	1.2	13
Ethiopia	0.1	1.6	3.4
India	2.3	1.2	6
Philippines	1.2	1.8	11
Ghana	0.8	9.3	13

* Compiled and computed from statistics of the UN World Health Organization.
  Statistics in this table are only approximate and roughly comparable. Counts were not all made at the same time. Dates were 1970 ± 2. See also Table 7.
† Data incomplete.

Table 25. U.S. Assistance and Military Outlays, 1945–1973*

Economic and Military Programs

Table A: U.S. government foreign grants and credits[a,b]

	1945–1973	1945–1955	1956–1965	1970	1971	1972	1973
Total, net[c]	155.6	55.1	47.3	5.70	7.13	7.94	6.77
Invested in five international agencies[d]	2.8	0.6	0.7	0.23	0.25	0.27	0.37
U.S. assistance program, net	152.8	54.5	46.7	5.46	6.88	7.67	6.39
Military grants, net[e]	58.9	16.4	19.2	2.55	3.24	4.22	2.82
Other grants, credits, assistance[f]	93.9	38.0	27.5	2.91	3.64	3.45	3.57
Europe[g]	25.5	25.8	−0.5	−0.27	−0.14	−0.08	0.35
Near East and Southern Asia[h]	25.5	3.1	12.3	0.99	1.61	1.10	1.03
Far East and Pacific[i]	22.8	6.8	7.8	0.98	0.98	1.22	0.94
Africa[k]	4.7	0.1	2.1	0.28	0.35	0.26	0.30
Western Hemisphere[m]	10.4	1.2	4.5	0.54	0.41	0.46	0.52
unspecified and miscellaneous	5.0	1.0	1.3	0.40	0.44	0.49	0.44

Table B: U.S. military outlays[a,n]

	1950	1960	1965	1970	1971	1972	1973
Budget outlays, defense, and veterans	21.9	51.3	55.3	89.0	87.4	89.1	88.0
National defense functions	13.1	45.9	49.6	80.3	77.7	78.3	76.0
Operating costs	—	10.2	12.3	21.6	20.9	21.7	21.1
Military personnel	—	11.7	14.8	25.9	24.8	26.9	27.6
Procurement	—	13.3	11.3	21.6	18.9	17.1	15.7
Research and development	—	4.7	6.2	7.2	7.3	7.9	8.2
Atomic energy program	—	2.6	2.6	2.5	2.3	2.4	2.4
Military assistance	—	1.6	1.1	0.7	1.0	0.8	0.5
Outlays for the three services							
Air Force	—	—	18.1	25.2	24.7	24.8	24.5
Army	—	—	11.6	25.1	23.9	23.5	21.1
Navy	—	—	13.3	22.7	22.4	22.7	23.0

Table 25. (Continued)

Table C: Arms exports by the U.S.[a,p]

	1961–1970	1971	1972	1973
U.S. exports of arms	19.42	3.38	4.26	5.02
World exports of arms	42.37	6.18	8.42	9.20

* Table A was compiled from the periodic publication of the BEA, *Foreign Grants and Credits by the U.S. Government.* Table B lists budget outlays as published by the Office of Management and Budget, except that division among the three services is from the Treasury Department. Table C from *The International Transfer of Conventional Arms,* and *World Military Expenditures and Arms Trade,* both published by the Arms Control and Disarmament Agency, 1973.

[a] Dollar quantities (in billions of current dollars) are given as originally published. See Table 29 for effects of inflation.

[b] A *grant* is a transfer for which no payment is expected. A *credit* assumes at least partial repayment.

[c] All repayments and returns have been deducted.

[d] Total investments in the five international agencies: Inter-American Development Bank, 41%; IDA, 32%; International Bank for Reconstruction and Development, 23%; Asian Development Bank, 3%; International Finance Corporation, 1%.

[e] About 54% of this total was for the Far East and the Pacific. The $58.9 billion is a small fraction of the total military outlay for 1945–1973 (see Table B).

[f] This is net economic and technical aid and credits. The *net* total of *grants* was $61.5 billion, of which two-thirds went to the developing countries. Of the gross total of grants, 17% was farm products and their transportation, and only 1.5% went to the Peace Corps. The *gross* total of *credits* was $55.1 billion, of which 36% was advanced to the Export-Import Bank.

[g] No net outlay since 1955. Largest net outlays to United Kingdom, 24%; France, 16%; Italy, 12%; West Germany, 11%; Yugoslavia, 8%.

[h] Includes Arab countries; Israel, 8%; Greece, 7%; Turkey, 10%; India, 35%; Pakistan, 17%.

[i] Includes Vietnam, 26%; Korea, 24%; Taiwan, 10%; Indonesia; Philippines.

[k] Largest outlays to Morocco, 17%; Tunisia, 15%; Zaire, 9%.

[m] Largest outlays to Brazil, 26%; Chile, 12%; Colombia, 11%.

[n] Official federal government outlays only.

[p] Total of military aid and subsidies is very difficult to determine. This table is included in order to provide a comparison with total arms exports. According to the Defense Security Assistance Agency of the Department of Defense, the "foreign military sales" of equipment, supplies, and services, authorized under the Foreign Military Sales Act and representing sales for cash or under government credit or guaranty, amounted to $14.79 billion for the period 1950–1973. According to the Agency for International Development, the total foreign military aid from the United States for 1946–1973 was $62.1 billion.

Table 26. Development Assistance for Developing Countries*

OECD and Other Official Programs

	U.S. current dollars (billions)				
	1968	1969	1970	1971	1972
Total outflow of resources for OECD development aid†					
Official direct, bilateral	—	5.37	5.58	6.26	6.53
Multilateral, World Bank, and so on	6.77	1.06	1.14	1.35	1.92
Miscellaneous official	—	0.58	1.12	1.13	1.51
Private direct, bilateral	4.80	5.46	5.91	6.73	7.09
Other private	0.61	0.42	0.47	0.77	0.67
Total	12.18	12.89	14.23	16.38	17.71
Financial terms of official development commitments, OECD‡					
Grants and distributions (%)	58	66	63	59	63
Loans (%)	42	34	37	41	37
Bilateral commitments by centrally planned economies (CPE)§					
China	—	—	0.69	0.49	0.34
USSR	—	0.40	0.20	0.68	0.66
East Germany	—	0.13	0.13	0.03	0.02
Romania	—	0.13	0.01	0.14	0.23
Bulgaria, Czechoslovakia, Hungary, Poland	—	0.12	0.23	0.17	0.24
Total	—	0.78	1.26	1.51	1.49

Official development assistance, by Region, 1970–1972 average‖

	OECD		CPE
	Direct	Multilateral	Direct
Asia, five countries	2.301	0.242	0.452
Asia, other countries	0.767	0.102	0.056
Africa	1.495	0.421	0.662
America	0.847	0.111	0.249
Oceania	0.298	0.007	0.000
Total	5.708	0.883	1.419

* Compiled from data published by the UN Center for Development Planning. From UN Statistical Yearbook 1973, sections 192–196, with permission.

† Statistics furnished by the Development Assistance Committee of the OECD. The amounts represent development assistance furnished by the developed marketing economies, the seventeen countries that are members of the OCED, namely, the United States and Canada, Western Europe (except Spain), Australia, New Zealand, and South Africa. The official development assistance is expressly for economic and social development.

‡ Covers the total official commitments of sixteen members of the OECD excluding New Zealand.

Table 26. (Continued)

§ Statistics collected by the UN for bilateral commitments of the centrally planned economies to twenty-eight of the developing countries.

‖ For OECD the five Asian countries receiving major assistance were India, South Vietnam, Indonesia, Pakistan, and South Korea; thirty-one other Asian countries received OECD assistance, as did forty-eight African countries, thirty-five American countries (Latin America), and many islands.

For the CPE, the five Asian countries receiving major assistance were Iraq, Iran, Pakistan, Afghanistan and Sri Lanka. CPE assistance went to four other Asian countries, thirteen African countries, and nine Latin American countries. Largest CPE commitments in Africa and Latin America were to Egypt, Algeria, Brazil, and Chile.

Table 27. World Status of Education, 1970–1971*

Teachers and Students per 10,000 Population

Area	Primary and preprimary	Secondary and high school	College and university	Total, all schools
Africa				
Teachers	23	7	1	31
Students	940	147	13	1100
Students per teacher	41	21	13	36
Latin America				
Teachers	26	16	3	45
Students	848	210	30	1088
Students per teacher	33	13	10	24
Asia†				
Teachers	31	14	3	48
Students	1068	279	42	1389
Students per teacher	34	20	14	29
Europe				
Teachers	49	43	7	99
Students	1136	621	101	1858
Students per teacher	23	14	14	19
Oceania				
Teachers	54	39	10	103
Students	1502	710	118	2330
Students per teacher	28	18	12	23
United States and Canada				
Teachers	60	50	33	143
Students	1450	960	404	2814
Students per teacher	24	19	12	20
USSR				
Teachers	Data not available			
Students	1450	377	189	2228

* Compiled and computed; UN sources.

   Averages for major world areas.

† China, North Korea, and North Vietnam not included.

Table 28.   U.S. Growth by Decades, 1900–1973*

Ten Economic Indicators

	Population increase in previous 10 years (%)[a]	Average annual birthrate per 1000 population[b]	Federal expenditures in year (dollars, billions)[c]	Foreign trade imports and exports in year (dollars, billions)[d]	Federal debt (dollars per capita)[e]
1973	11.2	15.0	246.	198.6	2177.
1970	13.4	18.3	195.	122.2	1811.
1960	18.6	23.8	77.2	51.9	1585.
1950	14.9	23.9	39.6	26.5	1697.
1940	7.7	19.4	9.1	9.0	367.
1930	15.6	21.3	3.4	9.9	131.
1920	15.3	27.7	6.4	17.0	228.
1910	21.4	30.1	0.7	4.3	12.
1900	21.0	32.3	0.5	2.9	17.

	Employment in farming (% of total employment)[f]	Employment in service occupations (% of total)[g]	Economic productivity per worker, 1970 = 100[h]	Telephones per 1000 population[i]	Inflation purchasing power of consumer's dollar, 1970 = 1.00[j]
1973	4.0	13.2	117	655	$0.87
1972	4.1	13.4	110	628	0.93
1971	4.2	—	103	601	0.96
1970	4.3	12.4	100	584	1.00
1965	6.0	12.6	86	478	1.23
1960	8.2	12.2	67	408	1.31
1955	10.4	11.3	60	—	1.45
1950	12.2	10.9	49	281	1.61

* See also Tables 4, 5, 10, and 38.
[a] Computed from Table 4.
[b] From Public Health Service.
[c] From Treasury Department.
[d] From Department of Commerce, BEA.
[e] From Treasury Department.
[f] From the Census: percent of employed civilian labor force.
[g] From Bureau of Labor Statistics (sales not included).
[h] From Bureau of Labor Statistics, private economy.
[i] From *Statistical Abstract of the United States,* 1974.
[j] From Table 29.

Table 29.   Inflation: The U.S. Dollar, 1900–1975*

Equivalent Values Expressed in 1970 Dollars

Table A: Constant dollars quoted in this book, based on the Wholesale
Price Index

Year	Factor	Year	Factor	Year	Factor	Year	Factor
1975†	0.63	1967	1.10	1955	1.26	1915	3.10
1974	0.70	1966	1.11	1950	1.35	1910	3.10
1973	0.81	1965	1.14	1945	2.02	1905	3.63
1972	0.92	1964	1.17	1940	2.73	1900	3.88
1971	0.97	1963	1.17	1935	2.70		
1970	1.00	1962	1.16	1930	2.50		
1969	1.04	1961	1.17	1925	2.09		
1968	1.08	1960	1.16	1920	1.40		

Table B: Constant dollars based on the Consumer Price Index

Year	Factor	Year	Factor	Year	Factor	Year	Factor
1975†	0.72	1970	1.00	1965	1.23	1960	1.31
1974	0.79	1969	1.06	1964	1.25	1955	1.45
1973	0.87	1968	1.12	1963	1.27	1950	1.61
1972	0.93	1967	1.16	1962	1.28	1945	2.16
1971	0.96	1966	1.20	1961	1.30	1940	2.77

* Values of the tabulated factors have been computed with a base of 1970 = 1.00, from sources as indicated in the explanation.

Explanation: The wholesale price index and the consumer Price Index are computed and published by the Bureau of Labor Statistics. There have been occasional changes in the methods of compilation. Since 1960 the Wholesale Price Index has been based on the large-quantity primary sales of almost 3000 commodities, weighted according to total net selling value. The Consumer Price Index is weighted to represent the typical expenditures of wage earners and clerical workers. Since 1964 it has been based on the prices of about 400 goods and services. Prices of the important items are collected monthly in 56 areas and weighted by population.

For specific purposes a special price index is often based on a limited group of commodities, and several of these are also published by the bureau. Sometimes the GNP Price Deflator is used, especially to show the difference in prices between private and government sectors. There is no official and satisfactory index of year-to-year inflation.

Throughout this book 1970 has been used as a base year for comparisons. Values expressed in constant dollars are based on the Wholesale Price Index of all commodities. Multiply earlier or later dollars by the factors given.

† 1975 data are preliminary and unofficial.

Table 30. U.S. Government Outlays for Designated Programs, 1960–1974*

Percentage Distribution of Federal Government Expenditures

Program or function	1960	1965	1970	1971	1972	1973	1974†
National defense and veterans benefits	55.7	46.7	45.2	41.3	38.4	35.7	34.1
Social security, unemployment benefits, disability assistance	19.7	21.7	22.3	26.4	28.0	29.7	30.9
Health, education, and manpower training	2.0	3.3	10.3	10.9	11.6	11.6	12.4
Commerce, transportation, communication	5.2	6.2	4.7	5.3	4.8	5.3	4.9
Agriculture and rural development	3.6	4.1	3.2	2.4	3.0	2.5	1.5
Community development and housing	1.1	0.2	1.5	1.6	1.8	1.7	2.0
Natural resources and environment	1.1	1.7	1.3	1.3	1.6	0.2	0.2
Space technology and research	0.4	4.3	1.9	1.6	1.5	1.3	1.2
Foreign aid, food and financial assistance	3.3	3.7	1.8	1.5	1.6	1.2	1.4
Interest on debt and miscellaneous	7.9	8.1	7.8	7.7	7.7	10.8§	11.4§
Total federal outlay (current dollars, billions)	92	118	197	211	232	247	270
Total federal outlay (1970 dollars, billions)‡	107	135	197	205	213	201	190

* Data from the Office of Management and Budget.
† Official estimate, late 1974. (Some classifications changed, 1975).
‡ Based on wholesale prices, Table 29.
§ Includes revenue sharing.

Table 31. Future Billions for New Projects*

New Funds Available for a 1970 Style of Life

	Annual amounts in 1970 dollars				
	1970	1980	1985	1990	2000
GNP for the year (dollars, billions)	977	1250	1415	1600	2050
Personal consumption (dollars, billions)	618	681	717	754	832
All other expenditures (dollars, billions)	359	396	417	438	484
Balance, for new projects (dollars, billions)	0	173	281	408	733
Total GNP, per capita (dollars)	4770	5530	5950	6405	7420
Personal consumption, per capita (dollars)	3015	3015	3015	3015	3015
All other expenses, per capita (dollars)	1755	1755	1755	1755	1755
Available for new projects, per capita (dollars)	0	760	1180	1635	2650

*Computed, see Table 33.

Explanation: Personal consumption spending in 1970-was $3015 per capita (Table 33), and total GNP was $4770 per capita, leaving $1755 per capita for all other expenditures. Keeping these per capita levels as "constant standards of living," while the GNP grows at 2½% per year, all the growth income would be available for new projects.

Assumptions: Slow growth of GNP at 2½% per year, and population growing 1% per year. Personal consumption per capita is constant.

Table 32.   Some U.S. Economic Comparisons, 1973 versus 1950*

Economic Activities in Current and Constant Dollars

Item	Apparent, based on current dollars (billions)			Corrected, based on constant 1970 dollars (billions)†		
	1950	1973	Ratio	1950	1973	Ratio
1. Population, total (millions)	—	—	—	152.3	210.4	1.38
2. Employed civilian labor force (millions)	—	—	—	58.9	84.4	1.41
3. Productivity index, output per work-hour, 1967 = 100						
Total private economy	—	—	—	60	115	1.92
Private nonfarm	—	—	—	65	113	1.74
Manufacturing only	—	—	—	64	127	1.98
4. GNP	285	1295	4.55	385	1049	2.72
5. GNP per capita (dollars)	1870	6155	3.29	2530	4980	1.96
6. Personal consumption	191	805	4.21	258	651	2.52
7. Private domestic investment	54	209	3.87	73	169	2.31
8. Personal saving	13.1	74.4‡	5.68	17.7	60.3	3.41
9. Federal government expenditures	42	255	6.0	57	207	3.64
10. State and local government expenditures§	23	181	7.9	31	147	4.74

* Sources: Items 1, 2, 8, 10: Bureau of the Census; item 3: Bureau of Labor Statistics; items 4, 5, 6: BEA, items 7, 9: Office of Management and Budget.
† Value of dollar based on average *wholesale* prices, all commodities.
‡ Revised data, 1975.
§ Direct general expenditures.

Table 33.   Distribution of the GNP, 1950–1973*

U.S. Statistics, Outgo Basis

	1950	1955	1960	1965	1970	1971	1972	1973
Total reported GNP	284.8	398.0	503.7	684.9	977.1	1055.	1155.	1289.
Personal consumption ($)	191.0	254.4	325.2	432.8	617.6	667.2	726.5	804.0
(%)	67.0	63.9	64.6	63.2	63.2	63.2	62.9	62.4
Durable goods ($)	30.5	39.6	45.3	66.3	91.3	103.6	117.4	130.8
(%)	10.7	9.9	9.0	9.7	9.3	9.8	10.2	10.1
Nondurables ($)	98.1	123.3	151.3	191.1	263.8	278.7	299.9	335.9
(%)	34.4	31.0	30.0	27.9	27.0	26.4	26.0	26.1
Services ($)	62.4	91.4	128.7	175.5	262.6	284.9	309.2	337.3
(%)	21.9	23.0	25.5	25.6	26.9	27.0	26.8	26.2
Private domestic investment ($)	54.1	67.4	74.8	108.1	136.3	153.2	178.3	202.1
(%)	19.0	16.9	14.8	15.8	13.9	14.5	15.4	15.7
Private residential ($)	19.4	23.3	22.8	27.2	31.2	42.7	54.0	58.0
(%)	6.8	5.9	4.5	4.0	3.2	4.0	4.7	4.5
Nonresidential ($)	27.9	38.1	48.4	71.3	100.6	104.4	118.2	136.2
(%)	9.8	9.6	9.6	10.4	10.3	9.9	10.2	10.5
Inventory added ($)	6.8	6.0	3.6	9.6	4.5	6.1	6.0	8.0
(%)	2.4	1.5	0.7	1.4	0.5	0.6	0.5	0.6
Government purchases ($)	37.9	74.2	99.6	137.0	219.5	234.3	255.0	277.1
(%)	13.3	18.6	19.8	20.0	22.5	22.2	22.1	21.5
National defense ($)	14.1	38.6	44.9	50.1	74.6	71.6	74.4	73.9
(%)	4.9	9.7	8.9	7.3	7.6	6.8	6.4	5.7
Other federal ($)	4.3	5.5	8.6	16.8	21.6	26.5	30.0	32.7
(%)	1.5	1.4	1.7	2.5	2.2	2.5	2.6	2.6
State and local ($)	19.5	30.1	46.1	70.1	123.3	136.2	150.5	170.5
(%) .	6.8	7.6	9.2	10.2	12.6	12.9	13.0	13.2
Net exports ($)	1.8	2.0	4.0	6.9	3.6	0.8	−4.6	5.8
(%)	0.6	0.5	0.8	1.0	0.4	0.1	− .4	0.4
Imports.($)	12.0	17.8	23.2	32.3	59.3	65.5	78.1	96.2
(%)	4.2	4.5	4.6	4.7	6.0	6.2	6.8	7.5
Exports ($)	13.8	19.8	27.2	39.2	62.9	66.3	73.5	102.0
(%)	4.8	5.0	5.4	5.7	6.4	6.3	6.4	7.9

* Data from the BEA.
Current dollars in billions and percentage distribution.

Table 34. Distribution of the GNP, 1950–1973*

U.S. Statistics, Income Basis

	1950	1955	1960	1965	1970	1971	1972	1973
Total reported GNP	284.8	398.0	503.7	684.9	977.1	1055.	1155.	1289.
GNP in 1970 dollars	385	501	585	781	977	1024	1119	1132
Personal income ($)	228.	311.	401.	539.	808.	864.	939.	1035.
(%)	80.0	78.1	79.6	78.7	82.7	81.9	81.3	80.3
Disposable income† ($)	207.	275.	350.	473.	692.	746.	797.	883.
(%)	72.7	69.1	69.5	69.0	70.8	70.7	69.0	68.5
Personal taxes, and transfers ($)	20.7	35.5	50.9	65.7	117.	118.	142.	153.
(%)	7.4	8.9	10.1	9.6	12.0	11.2	12.3	11.9
Business income ($)	87.0	122.8	161.1	227.7	311.7	345.5	380.8	434.1
(%)	30.5	30.8	32.0	33.2	31.9	32.7	33.0	33.7
Depreciation‡ ($)	18.3	31.5	43.4	59.8	87.3	93.8	102.	110.
(%)	6.3	7.9	8.6	8.7	8.9	8.9	8.8	8.5
Taxes and transfers ($)	24.1	33.3	47.1	65.2	97.5	107.	114.	123.
(%)	8.4	8.4	9.3	9.5	10.0	10.1	9.9	9.5
Social security ($)	6.9	11.1	20.7	26.6	57.7	64.6	73.7	92.1
(%)	2.5	2.8	4.1	3.9	5.9	6.1	6.4	7.1
Profits ($)	37.7	46.9	49.9	76.1	69.2	80.1	91.1	109.
(%)	13.3	11.8	9.9	11.1	7.1	7.6	7.9	8.5
Subtotal ($)	315.	433.8	562.1	766.7	1119.7	1209.5	1319.8	1469.1
(%)	110.6	108.9	111.6	111.9	114.5	114.6	114.3	114.0
Counted twice ($)	31.1	37.9	57.0	80.2	134.8	149.3	161.6	182.4
(%)	10.9	9.5	11.3	11.7	13.8	14.2	14.0	14.2
Transfer payments and interest§								
($)	22.3	27.4	43.6	60.4	110.1	124.2	135.6	154.6
(%)	7.8	6.9	8.7	8.8	11.3	11.8	11.7	12.0
Dividends ($)§	8.8	10.5	13.4	19.8	24.7	25.1	26.0	27.8
(%)	3.1	2.6	2.6	2.9	2.5	2.4	2.3	2.2

* Data from the BEA.

  Current dollars in billions and percentage distribution.

† Although total disposable personal income in current dollars continued to rise after 1973, the income per capita for 1973–1975, in constant dollars, did not increase appreciably.

‡ Capital consumption allowances.

§ These items were counted twice.

Table 35. Percentages of the GNP, 1950–1973*

Details of Incomes, Expenditures, and Investments

	1950	1955	1960	1965	1970	1971	1972	1973
GNP (current dollars, billions)	285	398	504	685	977	1056	1155	1289
Capital investments								
Entire economy	19.0	16.9	14.8	15.8	13.9	14.5	15.4	15.7
Manufacturing only	2.6	3.0	3.0	3.4	3.3	2.8	2.7	2.9
Utilities only	1.1	1.0	1.0	0.9	1.3	1.5	1.5	1.5
Corporation profits after taxes	8.7	6.8	5.3	6.8	4.0	4.5	4.8	5.5
Dividends to stockholders	3.1	2.6	2.7	2.9	2.5	2.4	2.3	2.2
Retained for investment	5.6	4.1	2.6	3.9	1.5	2.1	2.5	3.3
Profits in manufacturing only	4.6	3.8	3.0	4.0	2.9	2.9	3.2	3.7
New construction								
Total	11.8	11.7	10.9	10.8	9.7	10.4	10.7	10.5
All residential	6.5	5.6	4.7	4.2	3.4	4.2	4.8	4.5
All other, private	3.0	3.2	3.2	3.5	3.5	3.5	3.4	3.5
All other, public	2.3	2.9	3.0	3.1	2.8	2.7	2.5	2.5
Foreign trade								
Net exports	0.6	0.5	0.8	1.0	0.4	0.1	−0.4	0.4
Imports	4.2	4.5	4.6	4.7	6.0	6.2	6.8	7.5
Exports	4.8	5.0	5.4	5.7	6.4	6.3	6.4	7.9
Consumer finances†								
Consumer income	80.0	78.1	79.6	78.7	82.7	81.9	81.3	80.3
Taxes and transfers	7.4	8.9	10.1	9.6	12.0	11.2	12.3	11.9
Disposable income	72.6	69.1	69.5	69.0	70.8	70.7	69.0	68.5
Consumer spending	67.0	63.9	64.6	63.2	63.2	63.2	62.9	62.4
Durable goods	10.7	9.9	9.0	9.7	9.3	9.8	10.2	10.1
Nondurables	34.4	31.0	30.0	27.9	27.0	26.4	26.0	26.1
Services	21.9	23.0	25.5	25.6	26.9	27.4	26.8	26.2
Consumer saving	4.6	4.0	3.4	4.1	5.7	5.7	4.3	4.3
Government purchases								
Goods and services								
Total	18.5	21.4	18.8	16.7	14.3	13.1	12.4	11.2
Federal only	8.9	12.8	10.2	8.4	6.6	5.8	5.3	4.4
State and local	9.6	8.6	8.6	8.3	7.7	7.3	7.1	6.8
Federal budget outlays, total†								
All purposes	15	17	18	17	20	20	20	19
National defense	—	—	9.1	7.2	8.2	7.4	6.8	5.9
Income security‡	—	—	3.6	3.8	4.5	5.3	5.6	5.7

* Compiled and computed. Source: *Statistical Abstract of the United States, 1974.*
† All government tables show statistical discrepancies in GNP percentages when incomes and expenditures are mixed.
‡ Over two-thirds of this item is social security payments.

Table 36.   Performance of the U.S. Corporate System, 1950–1973*

Total Profits, Dividends, and Depreciation Allowances

	1950	1955	1960	1965	1969	1970	1971	1972	1973†
Profits before taxes[a]	42.6	48.6	49.7	77.8	84.9	74.0	85.1	98.0	127.
Income tax liability[b]	17.8	21.6	23.0	31.3	40.1	34.8	37.4	42.7	56.1
Profits after taxes	24.9	27.0	26.7	46.5	44.8	39.3	47.6	55.4	70.5
Net dividends[c]	8.8	10.5	13.4	19.8	24.3	24.7	25.1	26.0	27.8
Undisttributed profits	16.0	16.5	13.2	26.7	20.5	14.6	22.5	29.3	42.7
Capital consumption allowan-ces[d]	8.8	17.4	24.9	36.4	51.9	56.0	60.4	65.9	71.0
Cash flow[e]	24.8	33.9	38.2	63.1	72.4	70.5	82.9	95.2	114.
Profits plus allowances[f]	33.7	44.4	51.6	82.9	96.8	95.2	108.	121.	141.

* All corporations organized for profit, as reported by the BEA in *Survey of Current Business*.
   All values in current dollars (billions).
† 1973 statistics are preliminary.
[a] No deductions for depletion; exclusive of capital gains or losses; branch profits from abroad included.
[b] Federal and state.
[c] Paid to U.S. residents, after elimination of intercorporate dividends.
[d] Depreciation allowances and accidental damages.
[e] Undistributed profits plus capital consumption allowances.
[f] Net profits after taxes plus capital consumption allowances.

Table 37.   Average Growth of GNP and of Energy Supplies, 1970–1985*

Monetary Amounts in Constant 1970 Dollars

	Per Table 19	Estimates by others	
Real GNP per year			
Assumed annual growth (%)	2.5	3.0	4.0
Total GNP, 1970 (dollars, billions)†	977.	977.	977.
Total GNP, 1985 (dollars, billions)	1415.	1520.	1760.
Average GNP, 1970–1985	1197.	1248.	1356.
Total private investment, 15.5% of average GNP,			
1970–1985 (dollars, billions)	185.	193.	210.
Total energy supply per year			
Assumed annual growth (%)	2.5	3.0	4.0
Total energy 1970 (Btu $\times$ 10^{-15})‡	67.1	67.1	67.1
Total energy, 1985 (Btu $\times$ 10^{-15})	97.2	104.5	121.0
Electrical energy supply and plants			
Assumed annual growth (%)	4.5	5.0	7.0
Total electrical generation, 1970 (kWh $\times$ 10^{-12})§	1.53	1.53	1.53
Total electrical generation, 1985 (kWh $\times$ 10^{-12})	2.96	3.18	4.22
Total installed capacity, 1970 (1000 MW)‖	341.	341.	341.
Total installed capacity, 1985 (1000 MW)	660.	709.	941.
Capacity added, 1970–1985 (1000 MW)	319.	368.	600.
Cost of new capacity, 1970–1985 (dollars, billions)	160.	185.	300.
Capital required, average yearly, 1970–1985			
(dollars, billions)	10.7	12.3	20.0

* Compiled and computed from growth rates and references indicated.
† From Table 34.
‡ From Table 12.
§ From Table 14.
‖ From Table 14.

Table 38. Capital Investments by U.S. Nonenergy Industries, 1969–1974*

New Plant and Equipment (billions of Current dollars)

	1969	1970	1971	1972	1973	1974
1. All industries†	75.56	79.71	81.21	88.44	99.74	112.4
2. Nondurable goods manufacturing	10.09	10.54	9.98	10.46	13.32	15.38
3. Food and beverages	2.59	2.84	2.69	2.55	3.11	3.25
4. Paper	1.58	1.65	1.25	1.38	1.86	2.58
5. Chemical	3.10	3.44	3.44	3.45	4.46	5.69
6. Rubber	1.09	0.94	0.84	1.08	1.56	1.47
7. Other nondurable manufacturing	1.73	1.67	1.76	2.00	2.33	2.39
8. Railroads	1.86	1.78	1.67	1.80	1.96	2.54
9. Air and other transportation	4.19	4.26	3.26	3.92	4.07	4.12
10. Stone, clay, glass, and other nonmetal	4.51	4.40	4.30	4.07	5.45	6.54
11. Communication	8.30	10.10	10.77	11.89	12.85	13.96
12. Commercial and other	16.05	16.59	18.05	20.07	21.40	22.05
13. Total, items 3 to 12	45.00	47.67	48.03	52.21	59.05	64.59
14. Total, percentage of item 1	59.6	59.8	59.1	60.2	59.2	57.5

* Condensed from *Statistical Abstracts of the United States,* Table 811, 1975. Original source: BEA.
† Excludes agriculture, professions, institutions, and real estate firms.

Table 39. Growth of U.S. Manpower Since 1950*†

Trends That Will Affect the Future

(1)	(2)	(3)	(4)	(5)	(6)	(7)	(8)	(9)
	Popula-tion increase (thou-	Popula-tion growth in year	Birth-rate per 1000 popula-	Total civilian labor force (mil-	Total in service occupa-tions (mil-	College degrees in engineering (thousands)		Total college enroll-ment (thou-
Year	sands)	(%)	tion	lions)	lions)	Bachelor's	Advanced	sands)
1974	1504	0.71	14.9‡	91.01‡	—	41.4	19.2	8520
1973	1554	0.77	15.0	88.71	15.47	43.4	20.7	8270
1972	1797	0.87	15.6	86.54	14.80	44.2	21.1	8120
1971	2170	1.06	17.3	84.11	14.22	43.2	20.0	—
1970	2198	1.09	18.4	82.71	14.03	43.0	19.2	7480
1969	1971	0.98	17.8	80.73	13.75	40.0	18.3	—
1968	1994	1.01	17.5	78.74	13.27	38.0	18.1	6410
1967	2152	1.09	17.8	77.35	12.81	36.2	16.5	—
1966	2257	1.16	18.4	75.77	12.20	35.8	16.0	5530
1965	2414	1.24	19.4	74.46	11.76	36.7	14.2	—
1964	2647	1.40	21.0	73.09	11.00	35.2	12.8	4990
1963	2704	1.45	21.7	71.83	10.50	33.5	11.0	—
1962	2847	1.55	22.4	70.77	10.27	34.7	10.1	3730
1961	3020	1.67	23.3	70.70	10.23	35.9	9.1	—
1960	2841	1.60	23.7	69.82	10.17	37.8	7.9	3450
1959	2948	1.69	24.3	68.64	—	38.1	7.5	—
1958	2898	1.69	24.5	68.03	—	35.3	6.4	2900

Table 39. (Continued)

(1) Year	(2) Population increase (thousands)	(3) Population growth in year (%)	(4) Birth- rate per 1000 population	(5) Total civilian labor force (millions)	(6) Total in service occupations (millions)	(7) (8) College degrees in engineering (thousands) Bachelor's Advanced		(9) Total college enrollment (thousands)
1957	3081	1.82	25.2	66.95	—	31.2	5.8	—
1956	2972	1.79	25.2	66.56	—	26.3	5.3	2640
1955	2905	1.78	25.0	65.50	8.71§	22.6	5.1	—
1954	2842	1.77	25.3	64.47	—	22.2	4.8	2200
1953	2631	1.67	25.0	63.82	—	24.2	4.3	—
1952	2675	1.73	25.1	62.97	—	30.3	4.7	2300
1951	2607	1.71	24.9	62.88	—	41.9	5.7	—
1950	2504	1.67	24.1	62.60	7.98	50.7	5.4	3510⁪

* Sources of data (by column number):

  2. Bureau of the Census, "Current Population Reports," and "Historical Statistics."

  3. Computed from 2.

  4. Bureau of the Census and National Center for Health Statistics.

  5. Bureau of Labor Statistics.

  6. BEA.

  7 to 9. Office of Education and Engineering Manpower Commission.

† Explanations (by column number).

  2. Increase from July 1 to July 1, armed forces included.

  3. Increase as percentage of previous year's population; computed.

  4. Live births per 1000 population.

  5. Persons 16 years and older; civilians only.

  6. Equivalent full-time employment, work-years. Sales occupations are not included. Includes education, health, medical, and legal professions, business and repair services, nonprofit organizations, recreation, domestic service.

  7. To include technician degrees, add about 5% (recent years only).

  9. Degree-credit enrollment only; from biennial survey, Office of Education. For total enrollment, add about 10%. The undergraduate enrollment increased from about 30% of 18- to 21-year olds in 1950–1955 to over 40% in 1965–1974.

‡ Preliminary.

§ Years omitted due to classification changes; data unavailable.

ⁱⁱ Includes extension and correspondence, some nondegree.

Index